Remembering the Manhattan Project

Perspectives on the Making of the Atomic Bomb and its Legacy

Remembering the Manhattan Project

Perspectives on the Making of the Atomic Bomb and its Legacy

editor

Cynthia C. Kelly

President, The Atomic Heritage Foundation, USA

World Scientific

NEW JERSEY • LONDON • SINGAPORE • BEIJING • SHANGHAI • HONG KONG • TAIPEI • CHENNAI

Published by

World Scientific Publishing Co. Pte. Ltd.

5 Toh Tuck Link, Singapore 596224

USA office: 27 Warren Street, Suite 401-402, Hackensack, NJ 07601

UK office: 57 Shelton Street, Covent Garden, London WC2H 9HE

British Library Cataloguing-in-Publication Data
A catalogue record for this book is available from the British Library.

REMEMBERING THE MANHATTAN PROJECT
Proceedings of the Atomic Heritage Foundation's Symposium on the Manhattan Project

ISBN 981-256-040-8

Printed in Singapore.

PART I: A REPORT ON THE PROCEEDINGS OF THE ATOMIC HERITAGE FOUNDATION'S SYMPOSIUM ON THE MANHATTAN PROJECT

Saturday, April 27, 2002
Carnegie Institution of Washington
1530 P Street, N.W.
Washington, D.C.
U.S.A.

PART II: A PLAN FOR PRESERVING THE MANHATTAN PROJECT

Preserving America: A Strategy for the Manhattan Project
Interim report to Congress prepared by the
Atomic Heritage Foundation

CONTENTS

PART I: A REPORT ON THE PROCEEDINGS OF THE ATOMIC HERITAGE FOUNDATION'S SYMPOSIUM ON THE MANHATTAN PROJECT

Chapter 1: A History Worth Preserving

Opening Remarks

Senator Jeff Bingaman, New Mexico

Let me first say I'm honored to be here, particularly at a symposium like this that is attended by many who lived this history as well as many who have devoted their careers to writing it and bringing it alive for the American public and the whole world. This is an exciting subject and Cindy I complement you, as President of the Atomic Heritage Foundation, for organizing this and for the effort that is being made to put together the matching funds for the $700,000 challenge grant that was appropriated through the public-private partnership called Save America's Treasures, in order to help preserve the history of the Manhattan Project. There's obviously no more important and more fascinating chapter of American history — American science and engineering history, in particular. For someone like myself, who grew up a hundred miles west of the Trinity site in southwest New Mexico, this has been a subject of fascination for all my life: how this group of extraordinary individuals came together — scientists and engineers primarily, but also others. I know General Groves is going to be spoken about here at length as well, and there are many with a scientific and engineering background who came together to marshal the support of the political leadership of this country to put the country on the path to pursuing the Manhattan Project. It was pursued to a successful conclusion, the bomb was developed, the war was ended, and there are so many exciting aspects to this history that it's hard to know where to begin. I will leave most of the description of the history to the superb group of historians that we have here to speak today.

Let me just relate three of the stories that I've always enjoyed very much that have come out of this history, because they are things that I've encountered in the reading that I've done. The first is right from the book Richard Rhodes did, his extremely well-received book, "The Making of the Atomic Bomb." It talks about the difficult interface between the military establishment, or the military culture, and the scientific culture, and it involves of course Leo Szilard, and I'll just read a very small portion of it:

> "Somewhere along the way [General] Groves put Szilard under surveillance. The Brigadier still harbored the incredible notion that Leo Szilard might be a German agent. [...] The surveillance of an innocent but eccentric man makes gumshoe comedy. Szilard traveled to Washington on June 20, 1943, and in preparation for the visit an Army counterintelligence agent reviewed his file: *'The [...] Subject is of Jewish extraction, has a fondness for delicacies and frequently makes purchases in delicatessen stores, usually eats his breakfast in drug stores and other meals in restaurants, walks a great deal when he cannot secure a taxi, usually is shaved in a barber shop, speaks occasionally in a foreign tongue, and associates mostly with people of Jewish extraction. He is inclined to be rather absent-minded and eccentric, and will start out a door, turn around and come back, go out on the street without his coat or hat and frequently looks up and down the street as if he were watching for someone or did not know for sure where he wanted to go.'* "[1]

And it concludes by saying, "Armed with these profundities a Washington agent observed the Subject arriving at the Wardman Park Hotel at [...] 8:30 P.M. on June 20 [...]. Szilard worked the next morning at the Carnegie Institution with Captain Lavender."[2] So I think the history makes clear the appropriateness of the site that we are convened in today.

Let me go on to a little different aspect of the history and that is some of the history of the policy and politics that combined to commit the country to the Manhattan Project. I've always very much enjoyed the stories of Senator McKellar from Tennessee. Many of you may have heard this story, but it's one that my former colleague Senator Sasser liked to use very much when he was in the Senate. He's talking about how they persuaded Congress to appropriate the funds for the Manhattan Project.

> "[Secretary of War Henry Stimson] hinted in a meeting with several Senators that he needed McKellar to hide $2 billion in an appropriation

[1] Richard Rhodes, *The Making of the Atomic Bomb*, 1st ed. (New York: Simon & Schuster, 1988), pp. 506–7.
[2] Rhodes 507.

> bill for a secret project that might bring the war to an early end. That night, McKellar could not sleep. He returned to Stimson's office the next day to inquire further. 'Can you keep a secret?' Stimson asked. 'We are going to split the atom.' 'Here we are in the middle of a big war,' McKellar shot back, 'and you are fooling around, trying to split the atom.' Legend has it that McKellar took his concerns directly to the top and was actually with the President, who had summoned him to the White House to reiterate the request, when it finally dawned upon McKellar what Stimson had in mind. President Roosevelt asked, 'Senator McKellar, can you hide two billion dollars for this supersecret national defense project?' Senator McKellar, not missing a beat, replied, 'Well, Mr. President, of course I can. And where in Tennessee do you want me to hide it?' "[3]

And the third and final story that I will impose on you is one that Richard Feynman tells. Richard Feynman recounts some of the early problems with security at Los Alamos. That's a subject we've spent a lot of time on in recent years and it's not a new subject, so let me give you this short account of Richard Feynman's:

> "[...at] the very beginning we had terribly important secrets. We'd worked out lots of stuff about uranium, how it worked, and all this stuff was in documents that were in filing cabinets that were made out of wood that had on them little ordinary common padlocks. Various things made by the shop were on the cabinets, like a rod that would go down and then a padlock to hold it, but it was always just a padlock. Furthermore, you could get the stuff, without even opening the padlock out of these wooden cabinets: you just tilt it over backwards and from the bottom drawer you could extract the papers. Every time we had a meeting of the whole group [these were the meetings in Los Alamos, of course], every time we had a meeting of the whole group, and everybody together I would get up and I'd say that we have important secrets and we shouldn't keep them in such things. These were such poor locks, that we need better locks. And so one day Teller got up at the meeting and said to me,

[3] William Frist and J. Lee Annis, Jr., *Tennessee Senators, 1911–2001: Portraits of Leadership in a Century of Change* (Lanham, MD: Madison Books, 1999), pp. 29–30.

> "Well, I don't keep my most important secrets in my filing cabinet, I keep them in my desk drawer. Isn't that better?" I said "I don't know, I haven't seen your desk drawer." Well, he's sitting near the front of the meeting, I'm sitting further back. So the meeting continues and I sneak out of the meeting and I go down to see his desk drawer. I don't even have to pick the lock on the desk drawer. It turns out that if you put your hand in the back underneath you can pull the paper out like those toilet paper dispensers; you pull out one, it pulls another, it pulls another. I emptied the whole damn drawer, took everything out, and put it away to one side and then went up on the higher floor and came back. The meeting is just ending and everybody is just coming out I run up to catch up with Teller and say, "Oh, by the way, let me see your desk drawer. So he says, 'Certainly,' so we walk into his office and he shows me the desk and I look at it and say that looks pretty good to me. I said, "Let's see what you have in there." "I'd be very glad to show it to you," he says putting in the key and opening the drawer, "if you hadn't already seen it yourself." The trouble with playing a trick on a highly intelligent man like Mr. Teller is that the time it takes him to figure out from the moment that he sees there is something wrong till he understands exactly what happened is too damn small to give you any real pleasure!"[4]

This is history that each generation needs to know. I'm sure there are extremely important lessons for us in this history, and again I commend the Carnegie Institution for the symposium and I commend all of you for your role in this history, and also for your efforts to be sure that it's preserved for future generations. Thank you very much.

[4] Richard P. Feynman, *The Pleasure of Finding Things Out: The Best Short Works of Richard P. Feynman* (Cambridge, MA: Perseus Books, 1999), pp. 70–71.

Opening Remarks

Dr. Everet H. Beckner
Deputy Administrator for Defense Programs at the Department of Energy's National Nuclear Security Administration (NNSA)

Thank you very much. I am proud to be here representing Secretary Abraham of the Department of Energy and General (Ret.) John Gordon the Administrator for the National Nuclear Security Administration.

I thought it might be interesting to spend just a few minutes reminiscing on how I got involved in nuclear energy and indirectly in the Manhattan Project and the successor to the Manhattan Project. I too, like Senator Bingaman was born in northern New Mexico in a little town called Clayton, up in northeastern New Mexico. I fully remember the events at the end of the Second World War although I was too young to have done much more then wonder what was happening when the bombs were set off at Hiroshima and Nagasaki and the war ended.

I got my first glimpse of Los Alamos when we went up to play football, when I was a senior in high school; that would have been about 1952. The thing that I remember to this day as we were coming from northeastern New Mexico — I didn't really know where Los Alamos was at the time, certainly didn't know during the war — was security. But anyway we drove up in our yellow school bus, and I still remember that as we got to the outskirts of Los Alamos there was still a large gate out there that you had to be admitted through, a good distance outside of town. And there was an anti-aircraft gun standing beside it as well. You know this is more than six years after the war was over, but it was still there. I don't know what they planned to do with it but it was still there, so one way or another I associate that scene with my early experience with this program.

Subsequently, I went off to college, got involved in physics, and ended up I would say as an early beneficiary of the Atomic Energy Commission because I, like many others benefited from a small grant. I was in graduate school working in nuclear physics and a grant from the Atomic Energy Commission helped pay to run the accelerator in the laboratory that I worked in. So I guess that I would have to say that I have been associated with the program literally most of my adult life, and it's been a wonderful association.

It's hard to imagine how things might have gone otherwise, but they surely could have and I think that's what we should spend a bit of time thinking about today. Because I think it is really extraordinary that the effort that was put together during the war, for a very specific purpose, managed to survive the enormous changes that occurred following the war as the nation set about to find its way to peace, and at the same time to defend itself. The Atomic Energy Commission came into being, and the laboratory system came into being, and the whole complex then evolved under the Atomic Energy Commission, transitioning into the Energy Research and Development Agency in the 1970's as the county worked to find a way to become less dependent on energy from sources other than oil and gas; then following that, combining with other elements of energy programs and other agencies to become the Department of Energy. And so today you look around and I think we all will agree it's a magnificent set of capabilities that have been assembled. And I think in all fairness although I come to it from the science side, an enormous debt is owed to the people within the Congress and within the various Administrations over the years who have seen the importance of keeping it intact and of making it truly the institution or the set of institutions that we now have today. We have Senator Bingaman here today, as a representative of the Congress and those in the administration who have seen the importance of keeping these institutions alive.

Today, I think we would all agree, you could not do this again. You couldn't start it over. So the fact that it has survived, I think is enormously important, and most likely we will hear more about that from some of the other speakers today. The only other thing I can say at this point is that I don't have any way of seeing the future better than others of you, but I think it's fair to expect that these institutions will continue to be responsive to national needs. They responded in the 1970's to the energy crisis, and I think recently they have been very agile in turning their attention to threats in the world today having to do with terrorism. I think we'll see a lot of changes in the coming years.

That too is something that if you tried to manage it from Washington, you couldn't do it. But in fact, the element that was brought into the program early on was of having a set of contractors to manage these institutions with obviously very direct connections to the federal government. That has made a difference early on, which no one could have anticipated. Again, I don't think you could set it up and do it today but it survived after the war, continued for the fifty years since then. I would encourage us all to recognize the importance of that system and to be sure that we protect it over time because it gives us management in these institutions that brings the private sector into play in ways that really, I think to be very valuable.

So, I've done my bit of reminiscing at this point. I consider myself extremely fortunate to have been a part of the program now for more years than I ever imagined might occur. I recently had an opportunity to see and spend some time in the British program having just spent two years at the atomic weapons establishment in Aldermaston in England and things work differently there. I wouldn't say anything that was other than complimentary about their program, because it is a fine program. But they handle their program in a different way. As you get an opportunity to see the way the program there was managed, it does give you insight into the difference that some of those early decisions made. Decisions as to how this enterprise would go forward.

So all in all, I am happy to be here, I compliment all of you for taking the time to come. I am sure you are interested in the program. Thank you for your time.

Preserving the History of the Manhattan Project

Cynthia C. Kelly
President, Atomic Heritage Foundation

This Symposium on the Manhattan Project is an opportunity to learn about one of the most significant events in the 20th century, and to draw lessons from the Manhattan Project for meeting today's challenges. Each of the papers that follows provides a different perspective on what made the Manhattan Project succeed: to do in just 27 months what every other leading nation in World War II concluded was impossible.

While the development of the atomic bomb changed the course of world history, to date few of the original Manhattan Project structures have been preserved for the public and future generations. Most of these properties have not been accessible to the public, and in recent years have been stripped of their equipment and slated for demolition. Unless we act now to preserve some of the last remaining Manhattan Project properties, first-of-a-kind equipment, artifacts, and other tangible remnants of this remarkable undertaking will be lost.

For example, few people have seen the buildings where the Manhattan Project scientists, engineers, and hundreds of young military recruits worked in around-the-clock shifts to develop an atomic weapon at the "V Site" at Los Alamos. These very humble wooden structures are "monumental in their lack of monumentality," as San Francisco architect Bruce Judd described them. Three years ago, a cluster of Manhattan Project buildings "behind the fence" at Los Alamos called the "V Site" were awarded a Save America's Treasures grant so they might be restored for the public. Tragically, the Cerro Grande fire destroyed all but one of them in May 2000. The remaining "V Site" building pictured on the cover of this report was protected by its cladding of asbestos shingles.

If preserved, the reality of the original "V Site" and other Manhattan Project buildings and equipment at Los Alamos, Oak Ridge, and Hanford will help maintain critical elements of this remarkable history. As

Bulgarian-born Stephane Groueff observed, the Manhattan Project was a testament to the "American way," a combination of "readiness for risk-taking, courage for unorthodox approaches, serendipity, [and] dogged determination." The Manhattan Project buildings — some massive concrete structures and others no more than work sheds — will forcefully communicate this "American way," and the ingenuity and resourcefulness that were instrumental to winning the war.

The proceeds of this Symposium will help match the Save America's Treasures grant to restore the Manhattan Project properties at Los Alamos as an annex to the Bradbury Science Museum. If you would like to make a contribution to match the grant or learn more about the Atomic Heritage Foundation, please visit our website at www.atomicheritage.org or call 202-293-0045. Thank you for your interest and enjoy the Symposium on the Manhattan Project and the excellent papers included here.

Chapter 2: The Manhattan Project — A Millennial Transformation

THE ATOMIC BOMB IN THE SECOND WORLD WAR

Richard Rhodes
Independent journalist and author

Nuclear fission was discovered accidentally in Nazi Germany on December 21st, 1938, nine months before the beginning of the Second World War. It was a discovery that in the long run would sharply limit national sovereignty and change forever the relationship between nation-states, and it came as a complete surprise.

The German radiochemists Otto Hahn and Fritz Strassmann, working at the Kaiser Wilhelm Institute for Physical Chemistry in Dahlem, a suburb of Berlin, were bombarding a solution of uranium nitrate with lukewarm neutrons, transmuting microscopic quantities of the uranium into a brew of substances of differing characteristic radioactivities which the two chemists believed might be new manmade elements heavier than uranium as well as familiar elements like radium, one of uranium's natural "daughters." But instead of radium Hahn and Strassmann found barium in their irradiated solutions, an element about half as heavy as uranium that they had not expected to find there, and when they had consulted with their physicist colleague Lise Meitner, who was in exile in Sweden, and when she had consulted with her physicist nephew Otto Frisch, who came over from Denmark to visit her, it became clear that the unexpected barium was a marker for a new species of nuclear reaction, that in fact their neutrons had split uranium nuclei into two nearly equal pieces, one of barium and one of krypton, and that the new reaction was fiercely exothermic, ten million times as much energy coming out as the neutrons carried in. Physicists had known for forty years that enormous energy was locked up in the atom. Here at last was a way to release it. Otto Hahn brooded on the probable military applications of his discovery and seriously considered suicide.

Word spread quickly in the small world community of physicists. Hahn and Strassmann published their results in the German science journal *Naturwissenschaften*. Frisch and Meitner told Niels Bohr, the great Danish physicist, and followed up with confirming physical experiments published in the British journal *Nature*, borrowing from biology a name for the new reaction. Bohr brought the news to America. Soviet physicists working in Leningrad under young Igor Kurchatov, British physicists, German physicists, the French team at the Radium Institute in Paris, American experimenters from coast to coast rushed to demonstrate fission in their laboratories with off-the-shelf equipment; the discovery, as the physicist Philip Morrison would say much later, was "overripe."[5] A Japanese Army lieutenant general who was an electrical engineer assigned a member of his staff to track it. If Hahn and Strassmann hadn't discovered nuclear fission in Germany, others would have discovered it in some other laboratory somewhere else. Here was no Faustian bargain, as some even all these years later find it comforting to imagine. Here was no evil machinery one or another noble scientist might have hidden from the politicians and the generals. To the contrary, here was a new insight into how the world worked, an energetic reaction older than the earth that science had finally devised the instruments and arrangements to coax forth. "It is a profound and necessary truth," the American theoretical physicist Robert Oppenheimer would say, "that the deep things in science are not found because they are useful; they are found because it was possible to find them."

The physicists saw immediately what might be done with the new reaction. Hungarian émigré physicist Leo Szilard told his American patron Lewis Strauss on January 25, 1939, that nuclear energy might be a means of producing power and mentioned "atomic bombs."[6] Philip Morrison remembers that "when fission was discovered, within perhaps a week there was on the blackboard in Robert Oppenheimer's office [at Berkeley] a drawing — a very bad, an execrable drawing — of a bomb."[7] "These possibilities," comments American theoretician Robert Serber, "were

[5] Personal communication.

[6] Leo Szilard, *Leo Szilard: His Version of the Facts*, eds. Spencer Weart and Gertrud Weiss Szilard (Cambridge, MA: MIT Press, 1978), p. 62.

[7] Quoted in American Institute of Physics, Charles Weiner, ed., *Exploring the History of Nuclear Physics: Proceedings of the AIP and American Academy of Arts and Sciences Joint Conference Brookline, Mass.* (New York: Springer-Verlag, 1972), p. 90.

immediately obvious to any good physicist."[8] Within months of the German discovery, the Italian Nobel laureate Enrico Fermi would stand at his panoramic office window high in the physics tower at Columbia University, look down the gray winter length of Manhattan Island alive with vendors and taxis and crowds, cup his hands as if he were holding a ball and say simply, "A little bomb like that, and it would all disappear."[9]

Why would these men of good will, who believed themselves to be members of a peaceful international community of scientists, want to build a weapon of mass destruction? Always and everywhere in that first round of nuclear proliferation the same reason repeats: because possession of such a weapon seemed the only defense against an enemy similarly armed. Deterrence had already been debated publicly and at length during the 1930s in the context of aerial bombardment. It found its first documented expression in the context of nuclear weapons in a secret report prepared early in 1940 as a warning to the British government by two émigré physicists, Otto Frisch (again) and Rudolf Peierls, that also first laid out on paper the basic design and operation of an atomic bomb. "If one works on the assumption," the two physicists wrote, "that Germany is, or will be, in the possession of this weapon, it must be realized that no shelters are available that would be effective and could be used on a large scale. The most effective reply would be a counter-threat with a similar bomb. Therefore it seems to us important to start production as soon and as rapidly as possible, even if it is not intended to use the bomb as a means of attack."[10]

The world was at war. The new tool of nuclear energy, like all tools, might also serve as a weapon of War. Every major industrial nation began a program to build atomic bombs: the Germans, the British, the French before their surrender, the Soviets, the Americans, the Japanese. But nuclear weapons development required a massive commitment of government funds, funds that would have to be diverted from the

---

[8] Robert Serber, *The Los Alamos Primer*, ed. Richard Rhodes (Berkeley, CA: University of California Press, 1992), p. xxvii.

[9] Quoted in: Daniel J. Kevles, *The Physicists: The History of a Scientific Community in Modern America* (Vintage, 1979), p. 324.

[10] Serber, *op. cit.*, Appendix I: The Frisch–Peierls Memorandum, 82.

conventional prosecution of the war. If atomic bombs could be built they would be decisive, in which case no belligerent could afford *not* to pursue them. But making that judgment depended on two corollary assessments: the first, whether or not such weapons were inventable, whether nature would allow such an explosion to proceed; the second, whether or not the enemy was capable of inventing them in time to affect the outcome of the war. Both assessments depended critically on how much scientists trusted their governments and how much governments trusted their scientists.

Trust would not be a defining issue later, after the secret, the one and only secret — that the weapon worked — became known. This first time around, however, it was crucial, as the Russian physicist Victor Adamsky, who worked on the Soviet bomb, has pointed out:

> The tension [between the scientists and their governments,][Adamsky writes,] stemmed from the fact that there existed no *a priori* certainty of the possibility of creating an atomic bomb, and merely for clarification of the matter it was necessary to get through an interim stage: to create a device (the nuclear reactor) in order to perform a controlled chain reaction instead of the explosive kind. But the implementation of this stage requires tremendous expenses, incomparable to any of those previously spared for the benefit of scientific research. And it was necessary to tell this straight to your government, making it clear that the expenses may turn out to be in vain — an atomic bomb may not come out…. The American nuclear scientists…addressed the President...directly and described that complicated situation to him.... After a number of procrastinations which are inevitable even in a democratic society, a decision was taken in the USA to make the research as comprehensive as required by logic, disregarding the [un]certainty of the final result. …There was [no such confidence and mutual understanding] in Germany.[11]

[11] V. B. Adamsky, "Becoming a Citizen," *Andrei Sakharov: Facets of a Life*, ed. B. L. Altschuler *et al.* (Gif-sur-Yvette, France: Editions Frontieres, 1991), p. 26.

In the United States the trust was there, and President Franklin D. Roosevelt duly authorized a full-scale Anglo-American nuclear weapons program on October 9, 1941. In Germany the trust was not there on either side and the German program fragmented and stalled. After 1942 Werner Heisenberg, Otto Hahn, and Richard von Weizsächer turned their attention to building a nuclear reactor, and the bomb went by the board. Nor did the German scientists believe the Allies could do what they themselves had not judged feasible.[12]

The French program was stillborn. The Soviets, fighting for their lives against an almost overwhelming German invasion, put their early work on hold, not entirely convinced of its importance, and revived it in 1943 after the Red Army pushed back the Wehrmacht outside Moscow. The Japanese saw that a bomb program was beyond their resources, estimated incorrectly that it was also beyond American resources and scaled down their efforts to laboratory studies of uranium separation.

The Anglo-Americans knew very little of these developments. Until 1944, they raced against an imaginary German clock, calculating that from the discovery of fission forward, the Germans might have at least a two-year lead. Then another and more terrible clock ticked off the project's hours: the clock of the war itself, of the young men dying on the battlefields of Europe and Russia and the bloody Pacific beaches. The Germans, the British, the French had used poison gas in the last Great War, as they all said, to shorten the war and save lives; Robert Oppenheimer, recruiting scientists for a secret laboratory in New Mexico where the first bombs might be designed and built, whispered that he couldn't tell them what they would be doing but he could tell them that their work would end the war and save lives.

Before Oppenheimer began recruiting, Szilard, Fermi and their colleagues at Columbia and then at the University of Chicago had to accomplish the intermediate step Adamsky mentions: they had to build an experimental nuclear reactor to prove that it was possible to achieve a controlled chain reaction in uranium. This would be a slow-neutron chain reaction, multiplying in thousandths of a second and relatively easy

[12] Evidence to this point is abundant in the Farm Hall tapes.

to control, not the microsecond fast-neutron chain reaction that would proceed in a bomb, but fission was the source of the energy in both arrangements. The reactor design Szilard and Fermi worked out was a spherical assembly large as a two-car garage of graphite blocks drilled with blind holes into which would be inserted pucks or slugs of uranium oxide or metal. They needed 700,000 pounds of highly purified graphite; all the uranium metal they could get, which turned out to be 12,400 pounds; and some 80,000 pounds of oxide. None of these materials could be bought off the shelf; their manufacture had to be developed and subsidized.

The "secret" of the bomb would turn out to be industrial production on an enormous scale — 2.2 billion dollars worth by the end of the war in 1945 dollars, about the same as it cost twenty years later to go to the moon. What began as a tabletop experiment on a laboratory bench in Germany in 1938 became, in the United States, an industry comparable in scale to the United States automobile industry of the day.[13] Niels Bohr had gone back to Denmark in 1939 secure in the conviction that no nation could afford to build such an industry in time of war. The United States not only did so; it did so redundantly, pursuing three different and expensive paths to accumulating the necessary quantities of fissionable materials. The Manhattan Project, as the program to build atomic bombs was named, commanded a higher priority for materials and staff than any other program of the war — not because anyone thought the atomic bomb would win the war, but because its sole possession by an enemy might turn Allied victory abruptly into defeat.

Fermi called his construction a "pile" because it was made that way, by piling up layers of uranium-slugged graphite bricks crosswise one on top the other to achieve a critical mass. The pile would not only prove the chain reaction; it would also be a model cauldron for transmuting from uranium the first manmade element, plutonium, Glenn Seaborg's discovery, several times more fissionable than uranium itself. Fermi's pile went critical in a doubles squash court under the stands of Stagg Field at the University of Chicago on December 2, 1942, one year almost

[13] Bertrand Goldschmidt, *Atomic Adventure* (Pergamon, 1964), p. 35.

to the day after Pearl Harbor; by January, General Leslie R. Groves of the U.S. Army Corps of Engineers had acquired half a million acres of land on the Columbia River near the village of Hanford in eastern Washington State, and there in the next three years the Army would build a major complex of five-story graphite production reactors and remote-controlled plutonium separation plants the size of ocean liners. The huge separation plants were scaled up directly from Glenn Seaborg's microgram-quantity work that paralleled Fermi's at Chicago.

In the meantime, in Tennessee, General Groves had already begun construction on two other gigantic and radical factories. Plutonium, chemically different from uranium, could be separated from the uranium matrix in which it had bred, but plutonium was a new and unknown element, with unknown properties, and Groves had no intention of relying on it alone. Uranium, a metal known since 1789, would make a more conservative bomb.

Unfortunately for Groves, ordinary natural uranium, though it could be coaxed with slow neutrons into generating power, as Fermi had shown, wouldn't chain-react with fast neutrons and therefore wasn't suitable for a bomb. Natural uranium is a mixture of two variant physical forms, called isotopes. Only the rarer of the two forms, designated uranium-235, explosively chain-reacts. Only one part in 140 of natural uranium is U235. The rest is U238. That slight difference in mass, three units out of several hundred, was the only difference physicists could exploit to separate the two isotopes.

Of a number of separation systems explored in the early days of the Manhattan Project, two proved promising enough to develop at industrial scale. One involved diffusing gaseous uranium through a porous barrier. Lighter U235 diffused through slightly faster than heavier U238; by repeating the process several thousand times in a cascade of stages similar to the pipes and towers of an oil refinery, it proved possible to enrich the U235 portion of natural uranium to above 90 percent. The gaseous diffusion plant that Groves built at Oak Ridge, Tennessee, was a U-shaped structure half a mile long, the most highly automated plant of its day.

The other separation system was even more prodigal; it involved vaporizing natural uranium in a vacuum and pulling the resulting ionized gas through a powerful magnetic field — separating natural uranium into its component isotopes atom by atom. The lighter U235 would follow a tighter arc through the field than the heavier U238, and collectors around the bend could catch the partly-separated sputter. The machines built at Oak Ridge drew enough electricity to light a city, but the process was relatively efficient. During the war gaseous diffusion and electromagnetic separation worked in harness, enriched product from one system feeding to the other. By the summer of 1945, these gigantic factories had produced only some fifty kilograms of weapons-grade U235, about three critical masses, enough for one conservatively-designed bomb. Army intelligence agents carried the monthly Oak Ridge output to Oppenheimer's secret New Mexican laboratory in a briefcase by public train. Plutonium from Hanford convoyed to New Mexico in an ordinary Army ambulance.

As there were two kinds of nuclear materials, so there were two kinds of bomb. The more conservative design, nicknamed Little Boy, was a six-foot cannon with a ring of U235 fitted around the muzzle and a second piece fired up the barrel at the appropriate time like an artillery shell. When the two parts mated they formed a critical mass and exploded. The gun-type bomb was grossly inefficient, but experiments with diluted uranium confirmed that it was reliable, so much so that it was certified for use without proof testing. In any case there wasn't enough U235 for a test; the first Little Boy built was the one that was used.

Los Alamos, Oppenheimer's secret laboratory on a mesa northwest of Santa Fe, discovered to its horror in the spring of 1944 that impurities in reactor-bred plutonium made that material so unstable that a piece fired up a gun barrel even at 3,000 feet per second would melt before it had time to mate and explode. Oppenheimer ordered a massive shift of laboratory priorities to develop an alternative method of assembling a critical mass, a method the physicists named implosion: using a lensed sphere of high explosives to squeeze a subcritical sphere of plutonium to critical geometry and density. The laboratory worked night and day for the rest of 1944 and the first half of 1945 to develop implosion; the technology was sufficiently unreliable even then to require a full-scale test. The test, in the desert northwest of Alamogordo, counted down to zero just before dawn on July 16, 1945, the first full-scale, manmade fast-neutron chain reaction exploding with a force equivalent to that of 18,000 pounds of TNT, a great fireball lighting up the night, thrusting into the pristine desert air on a stem of roiling gas and smoke. "No one who saw it could forget it," said the director of the test, "a foul and awesome display."[14] That same morning, the destroyer *Indianapolis* steamed out of San Francisco bay carrying Little Boy and its uranium bullet, bound for Tinian Island in the Marianas, where the B-29s that would carry the weapons to Japan had staged out in June and July. Little Boy's target ring assembly followed by air on July 26. So did two high-explosive assemblies for plutonium bombs — the spherical implosion weapon nicknamed Fat Man, to be used after Little Boy, and a second Fat Man for which a plutonium core would be ready on August 12th. More atomic bombs would be ready if needed, Oppenheimer projected in late July: "...from possibly three in September to we hope seven or more in December."[15]

President Harry Truman was waiting eagerly at the Potsdam Conference for word of a successful test. It bucked him up. The Soviet Union was still officially neutral in the Pacific War; Stalin had promised to begin fighting the Japanese on August 15, and until the news came of the successful New Mexican test, Truman's major concern had been to shore up Stalin's commitment. The test changed the stakes; now Truman wanted to end the war *before* the Russians joined it, to exclude them from the victory and a share of the spoils. "Believe Japs will fold up before Russia comes in," he confided to his diary. "I am sure they will when Manhattan appears over their homeland."[16] General George Marshall, the Army Chief of Staff, whose judgment everyone respected, remembered later why there was not more discussion in those final days of summer about demonstrating or pocketing the bomb:

> "We regarded the matter of dropping the [atomic] bomb as exceedingly important [Marshall said]. We had just gone through a bitter experience at Okinawa [the last major island campaign, when the United States lost

[14] Kenneth Bainbridge, *All in Our Time*, ed. Jane Wilson (*Bulletin of the Atomic Scientists*, 1975), p. 230.

[15] U.S. National Archives, Manhattan Engineer District papers, 5E, *Terminal cables*.

[16] Quoted in: Robert H. Ferrell, ed., "Truman at Potsdam," *American Heritage* June–July 1980: 42.

> more than 12,500 men killed and missing and Japan more than 100,000 killed in eighty-two grim days of fighting]. This had been preceded by a number of similar experiences in other Pacific islands, north of Australia. The Japanese had demonstrated in each case [Marshall goes on] [that] they would not surrender and would fight to the death.... It was expected that resistance in Japan, with their home ties, would be even more severe. We had had the one hundred thousand people killed in Tokyo in one night of [conventional fire-] bombs, and it had had seemingly no effect whatsoever. It destroyed the Japanese cities, yes, but their morale was not affected as far as we could tell, not at all. So it seemed quite necessary, if we could, to shock them into action.... We had to end the war [Marshall concludes], we had to save American lives."[17]

And in truth the first atomic bombs would not be even a quantitative extension of the destruction strategic bombing had already wreaked on the cities of Japan; since late April, Curtis LeMay's B-29s had been systematically firebombing Japanese cities one by one to utter destruction, killing hundreds of thousands of civilians in the process; by August 1 the B-29s were burning down cities of less than 50,000 population, almost the only cities left in Japan to burn. Hiroshima and Nagasaki survived to be atomic-bombed only because they had been deliberately reserved.

The devastation of Hiroshima on August 6th was complete. Little Boy yielded 13.5 kilotons. Of 76,000 buildings in the city, 70,000 were damaged or destroyed, 48,000 totally. Ninety percent of all Hiroshima medical personnel were killed or disabled. Up to September 1st at least 70,000 people died. More died later of the effects of radiation.

George Marshall remembered that he was surprised and shocked that the Japanese didn't immediately sue for peace. "What we did not take into account," he said, "...was that the destruction would be so complete that it would be an appreciable time before the actual facts of the case would get to Tokyo.... There was no communication for at least a day, I think, and maybe longer."[18] The Air Force distributed millions of leaflets

[17] Quoted in: Leonard Mosley, *Marshall: Hero for Our Times* (Hearst Books: 1982) p. 337.

[18] Quoted in: Mosley, *op. cit.*, 340.

over Japanese cities in the next several days suggesting that skeptics "make inquiry as to what happened to Hiroshima" and asking the Japanese people to "petition the Emperor to end the war."[19] Conventional bombing continued as well.

But there was now a struggle for power in the Japanese government, no surrender emerged, and on August 9th Fat Man exploded over Nagasaki with a 22 kiloton yield, killing at least another 40,000 people and devastating another Japanese city. Finally, breaking tradition, Emperor Hirohito demanded that the government communicate its surrender, and reluctantly it did. In his historic broadcast to his people on August 15, Hirohito specifically cited "a new and most cruel bomb, the power of which to do damage is indeed incalculable, taking the toll of many innocent lives" as "the reason why We have ordered the acceptance of the provisions of the Joint Declaration of the Powers...."[20]

The atomic bombs exploded over Hiroshima and Nagasaki didn't *win* the Pacific war, but without question they *ended* the war. What might have followed had the war continued, no one can say with certainty, except that Hiroshima and Nagasaki would certainly have been firebombed, probably to equivalent loss of life. The Japanese might have surrendered. Or the Allies might have had to invade the Japanese home islands, as they were vigorously preparing to do. The Russians would have joined in that invasion, and having done so would certainly have insisted on a larger share of the spoils than the Kuril Islands. Japan might have been partitioned as Korea and Germany were partitioned.

Was it necessary to drop the bombs? Were they weapons of mass destruction? Was their use a crime against humanity? I think such questions as these beg the real question, which is why war became so much more destructive in the first half of the 20th century than ever before in the history of our species. It did so, I believe, because efficient killing technologies made the traditional exercise of national sovereignty pathological. According to the *United States Strategic Bombing Survey*,

---

[19] J. F. Monyahan to L. R. Groves, 23 May 1946, MED 314.7, History.

[20] Quoted in Herbert Feis, *The Atomic Bomb and the End of World War II* (Princeton, NJ: Princeton University Press, 1966), p. 248.

> "The number of civilian deaths in Japan greatly *exceeded* the number of strictly military deaths inflicted on the Japanese in combat by the armed forces of the United States. This statement is pregnant with significance, for if there still be doubt that the emphasis in warfare has shifted from military forces to the civilian population, then this fact should dissipate all uncertainty."[21]

With disease, blockade, famine and fire it had sometimes been possible to kill large numbers of people in the past; the 20th century arsenal of automatic weapons, massed artillery, poison gas and firebombing made such slaughter more certain and more efficient — made it industrial, mass-produced it, so that the number of deaths became a direct function of time and resources invested in the work, death cranked out like sausages or cars. The atomic bomb was the culmination of that process, a mechanism that visited total death upon its targets cheaply, indiscriminately and almost instantaneously; whether or not people died at Hiroshima and Nagasaki depended not on their identities — whether combatants or noncombatants, Korean forced laborers, American prisoners of war, pregnant women, children, grandmothers, newborn babies or Shinto priests — but merely on the accident of their distance from ground zero that day.

The closing days of the Second World War mark a turning point in human history, the point of entry into a new era when humankind for the first time acquired the means of its own destruction. Niels Bohr liked to say that the goal of science is not universal truth. Rather, Bohr thought, the modest but relentless goal of science is "the gradual removal of prejudices."[22] The discovery that the earth revolves around the sun removed the prejudice that the earth is the center of the universe. The discovery of microbes removed the prejudice that disease is a punishment from God. The discovery of evolution removed the prejudice that *Homo sapiens* is a separate and special creation. The discovery of how to release nuclear energy, and its application to build weapons of mass

[21] *United States Strategic Bombing Survey*, Vol. X (Garland, 1976), p. 2.
[22] Niels Bohr, *Atomic Physics and Human Knowledge* (John Wiley, 1958), p. 31.

destruction, is gradually removing the prejudice on which war itself is based: the insupportable conviction that there is a limited amount of energy available in the world to concentrate into explosives, that it is possible to accumulate more of such energy than one's enemies and thereby militarily to prevail.

So cheap, so portable, so holocaustal did nuclear weapons eventually become that even nation-states as belligerent as the United States and the Soviet Union preferred to sacrifice a portion of their sovereignty — preferred to forego the power to make total war — rather than be destroyed in their fury. Lesser wars continue, and will continue until the world community is sufficiently impressed with their destructive futility to forge new instruments of protection and new forms of citizenship. But world-scale war at least has been revealed to be historical, not universal, a manifestation of destructive technologies of limited scale. In the long history of human slaughter that is no small achievement. These are hard truths, but total war is harder. "It is still an unending source of surprise for me," the Polish mathematician Stanislaw Ulam comments in his autobiography, "to see how a few scribbles on a blackboard or on a sheet of paper could change the course of human affairs."[23]

[23] S. M. Ulam, *Adventures of a Mathematician* (New York: Charles Scribner's Sons, 1976), p. 5.

The Manhattan Project: An Extraordinary Achievement of the "American Way"

Stephane Groueff
Bulgarian-born journalist and author

To better illustrate any thesis or argument, it usually helps to choose striking, sometimes even extreme examples. It makes the discussion clearer and accessible to a larger public, to interested non-specialists. If this also applies when assessing the relationship between science, industry, and government, or the lessons we can learn from World War II, I cannot think of any more instructive and appropriate subject than the Manhattan Project.

Talking about laymen and non-specialists, just look at me, a foreign correspondent with no scientific education and no government background! The reason I became involved, years ago, and fascinated with the subject, was the extraordinary, almost apocalyptic, scope of the Project, which not only decided the outcome of the war in the Pacific, but also affected the lives of all of us. It was natural for a journalist to find myself greatly intrigued by the story of the building, in total secrecy, of the first atomic bomb. I could not understand why so little had been written and known about the fantastic obstacles surmounted during its construction and the Herculean endeavor it had required.

It was 20 years after Hiroshima, and very little, if anything, was known about the production of the bomb ingredients and the actual building of the weapon. And yet most experts had declared these feats as infeasible. I am not referring to the scientific aspect, already pretty well known, but about the engineering, manufacturing, and organizational aspects. So little was known about the solving of the supposedly unsolvable problems of this titanic enterprise. Secrecy was, of course, one explanation.

After many interviews, my first journalistic assessment of the magnitude of the Manhattan Project was largely confirmed. In the opinion of most engineers, scientists, industrialists, and historians, the grandiose wartime undertaking was among the greatest achievements of organized human effort in history, comparable to the building of the Pyramids and the Great Wall in China. Though such superlatives are usually exaggerated, a further study of the Manhattan Project convinced me that the scope of the performance was indeed without precedent.

I was surprised that the American public itself was still unaware of the prodigious adventure into which their country's industrial power had been launched, secretly and boldly, at a colossal cost and with unprecedented effort to build the bomb. It's the story of the determination of men who did not fully suspect the incredible difficulties of the task or the immense difference between problems of the laboratory and problems of actual large-scale production.

I saw it as a superb illustration of the way the American system operated in the 1940s, the manner in which America reacted to a concrete danger, to a precise task. As you know, the enormous majority of the people involved were not informed about the final purpose of their effort. For them, there was a gigantic challenge to be met, an apparently impossible job to be done, and they carried it out to an end splendidly.

By "The System," I do not mean only the government, the military and political organizations, the scientific and industrial establishments. I have in mind a specifically American spirit of the time, or call it "the American way" — the mentality, habits and culture of the individual citizens. I am talking of the unusual role in problem solving played by ingenuity, readiness for risk-taking, courage for unorthodox approaches, serendipity, or dogged determination.

Otherwise, would we have witnessed improbable episodes such as, for example, the secret, unauthorized use of 6,000 tons of silver from the U.S. Treasury Department for magnetic coils and electric conductors as substitute for unavailable copper? Would the DuPont Company risk its reputation and probably its very existence by blindly committing itself to huge production of an unknown, never-seen metal, Plutonium, just because the government had asked for it, without revealing any details? DuPont experts' first reaction was: "The entire project seems beyond human capability."

The architectural firm of Skidmore, Owings, and Merrill took a similar risk, accepting a request to start building from scratch an entire city, Oak Ridge, which finally reached a population of 70,000, without being given any specifications or right to ask questions. Other secret towns — Hanford and Los Alamos — were built in a similar way. What other society could have produced tough, independent-minded industry captains, like K.T. Keller of Chrysler, Dobbie Keith of Kellex, and dozens of others, who took gamblers' chances with their companies, once General Groves and his people tickled their patriotic chords?

And shall I mention the countless individual stories? Many sound like paradoxical anecdotes. The house painter Edward Norris, whose spray gun unwittingly contributed, at a desperate hour, to the solution of one of the most difficult problems of the entire project, the gaseous diffusion "barrier." Before he realized what hit him, Norris found himself co-leader of Dr. Edward Adler's laboratory at Columbia University, with several Ph.D. assistants. No less colorful was the mysterious use in 1943 of the label printing machine of the popular "Chicklets" chewing gum, at the "American Chickle" company. And what about the unlikely drafting of Leonard Shazkin, a printing expert at the *House Beautiful* magazine, who hadn't even heard of nuclear physics?

As for the remarkable organizational success, we can't omit what I would characterize as the "West Point qualities" of men like General Groves and his team. Does our society still produce such fearless, totally dedicated, never doubting, and also competent, officers? Judging by the recent Desert War and Afghanistan, I'm inclined to be optimistic.

On the political aspect of wartime America I would quote the readiness with which Harry Truman, then a Senator, and other Congressmen irritated by the Government's reluctance to answer questions about rumors of huge cities being built in secrecy, accepted the word of honor of Secretary of War Stimson, Vannevar Bush, and General Marshall that national interest required silence and abstained from further inquiring.

The question comes sometimes whether anything of the Manhattan Project's magnitude would be possible today. Groves told me many times that it would not be possible in peacetime. He believed firmly that many urgent decisions he had to take could not be made by committees, or by consent, or by asking permission each time. The unprecedented secrecy of the Project would be unthinkable in a democratic society, except maybe in a war. Groves was fully aware that in case of failure, he would have spent the rest of his life at Congressional investigations, for having spent $2 billion without the taxpayers' knowledge. He was saying this 37 years ago. I think that today a Manhattan Project would be even less possible.

Because this symposium honors Dr. Vannevar Bush as a splendid model for liaison between government and science, I would like to share with you a question that has bothered me, the outsider, for a long time. Maybe some of you insiders have an answer.

Since the Manhattan Project's inception in 1942, only a handful of men had full knowledge of its super-secret purpose and could exercise the supreme control over the construction of the bomb. They were President Roosevelt, Vice-President Henry Wallace, War Secretary Henry Stimson, and General George Marshall, the Chief of Staff. These four leaders, together with Vannevar Bush, F.D.R.'s principal liaison with the scientific community and his deputy, Dr. James Conant, formed the so-called "Top Policy Group," responsible, at least on paper, for all major decisions on the bomb development. I was surprised to learn that, in fact, this group *never* had a formal meeting! They had delegated the day-to-day management of the Project to General Groves, who wasted no time in taking over the entire operation and making most decisions concerning the production process.

To supervise Groves's performance, the Top Policy Group relied on a special four member Military Policy Committee, whose members were Bush, Conant, Admiral W. Purnell, and General Wilhelm Styer, and which met only a few times. General Marshall's contacts with Groves were infrequent and only on concrete problems requiring his personal attention. Almost as removed as the President and Stimson, Marshall did not keep in direct touch with the Manhattan Project's command, except at the very end, when the bomb was almost ready. These leaders were briefed about the progress only by the loosely structured Military Policy Committee — not very often, nor in any detail — and they didn't seem to show any particular eagerness to learn more.

As a writer, I was surprised to find out that, contrary to what one would assume today, and despite the immense consequences the success or failure of the Project was going to have on the future of humanity, the A-bomb did not seem to be among the top concerns in the mind of the nation's leaders. I was shocked to discover F.D.R.'s rather casual interest and involvement in the progress of the work, and not to find at the White House any sense of utmost urgency, impatience, and absolute priority that, I thought, the bomb construction should warrant.

When I asked General Groves whether he felt urgent pressure coming from the White House, he told me: "I don't know whether Roosevelt ever inquired about the progress. But Vannevar Bush saw him from time to time and kept him posted. Bush also talked to Harry Hopkins at times, but he didn't know any of the details. No pressure came from the War Department either. They all knew it was a big job that would take time to do. Marshall had a natural disposition not to interfere with subordinates. The President, Stimson, and Marshall did not keep direct contact with me, they depended entirely on the Military Policy Committee."

I asked: "How often did you report to General Marshall?"

"The first time was when I took the report to the President from the "Military Policy Committee," early in December 1942. The next report was in August 1943; I didn't take that over."

I asked: "So no report for nine months, and Marshall did *not* ask?"

"No. I may have taken an occasional memorandum over to him, when I wanted him to do something. All he knew was that I'd been assigned to it and he assumed that I was doing my best. It was very convenient from my standpoint, and it worked very well, but if I hadn't succeeded, it would have been a terrible affair, and I don't think anyone in his position should have allowed me to go freewheeling as it were. Now, I did see from time to time on different matters Harvey Bundy, Stimson's assistant, and he no doubt told his boss. But after my August report, there were only two more written reports, that's all.

"At that time, the atomic bomb wasn't an element of the war strategy. Marshall's idea was that we would fight the war without it, and if it came, why, then we would be prepared to use it. The bomb may not work, one shouldn't count on it as a sure thing. I started seeing Stimson only in the fall of '44. Then by April '45 I saw him all the time."

"But for two years you wouldn't see Stimson, or Marshall, or the President?" I inquired.

"No. I only saw President Roosevelt once."

I found this hard to believe. "But General," I said, "the public imagines that while America was preparing the new secret weapon with which to end the war, the men in charge — Roosevelt, Stimson, Marshall — lived in an almost unbearable suspense, waiting impatiently for news, constantly asking, "Is it ready?" and "How's it going?" eager, pressing. And you're telling me that for two years they didn't inquire and you didn't report, every day?"

"Remember that Bush was talking to the President from time to time," Groves said, "and undoubtedly each time he told him about this. But he didn't go to see him on this particularly."

I talked to Vannevar Bush. He told me that regular White House conferences on the progress of the bomb never existed. In fact, there had never been *any* conference at Roosevelt's or Stimson's level, devoted specifically to the bomb construction! Bush personally briefed the President from time to time, in a very general way. These private meetings at the White House lasted about 15 minutes and covered several topics, of which the A-bomb was not necessarily the first, nor the most urgent.

Important people were constantly competing for F.D.R.'s time and attention, and while Bush eagerly counted on each one of his precious 15 minutes, the President sometimes wasted time in passionate dissertation on some completely different problem. Once Bush had to listen in deep frustration to a long tirade against General de Gaulle. Another time, Mrs. Roosevelt entered unexpectedly, greeted the visitor with a friendly "Hi, Van!", gave a kiss to the President and asked to join them. The scientific briefing continued, with Eleanor interrupting from time to time to ask a question, with F.D.R. patiently and lovingly trying to explain.

Usually, Bush would bring a summary of the major problems requiring immediate action, with proposals of the course to be taken. Very often, but not always, the Manhattan Project's current problems would figure on this memo. The President would glance at it, sometimes asking a few questions, sometimes not, then would scribble his "O.K.-F.D.R." on the paper.

"Were there occasions when Roosevelt would disagree or question the suggestions?" I asked Bush. "No, never."

How far this was from the stereotype ideas about history-making, I couldn't help thinking. The successful building of the bomb was less due to official decision-makers, than to the real "doers," Groves and his men. I realized also that the General was never given the extraordinary powers he used; he simply took them. Each time precise instructions were missing, he didn't ask. He acted.

Chapter 3: The Allies and the Atomic Bomb

A Tale of Two Documents

Andrew Brown
Oncologist and historian of science

The Manhattan Project had its scientific gestation in England during 1940 and 1941. The leading figures in the work were either men who had been trained by Ernest Rutherford (1871–1937) at the Cavendish Laboratory in Cambridge or scientists who had sought refuge in England after Hitler came to power in Germany. Their contributions were encapsulated in two documents — the Frisch–Peierls Memorandum and the Maud Report — both a few pages long.

Sir James Chadwick with Rudolph Peierls (*left*) and G. I. Taylor (*right*) in 1974.

When Sir Ernest Rutherford became head of the Cavendish Laboratory at Cambridge University after the First World War, he was already the most renowned experimental physicist of his day. He had led the way in exploring the phenomena of natural radioactivity after the Curies' original discovery at the end of the nineteenth century; then in 1911 he had been first to propose in detail a planetary model of the atom. His idea was refined by Niels Bohr, and the now familiar image of a small central nucleus with *electrons* whizzing around it in fixed elliptical orbits took shape. Rutherford's goal from that time was to discover whatever he could about the nature of the infinitesimal nucleus. He continued laboratory experiments during the First World War, in between advising the Admiralty on submarine detection devices, and in 1917 achieved the first "splitting of the atom" — an event that he told the Admiralty was of more significance than the war itself. His move from Manchester University to Cambridge coincided with the return of many young scientists from wartime service; there was no shortage of talented recruits to the Cavendish.

In 1920 he gave an invited lecture at the Royal Society in London, where he suggested for the first time that there might be neutral particles as constituents of the atomic nucleus. Until then, the nucleus was thought to contain positively charged *protons* and negatively charged electrons: Rutherford's conception of the *neutron* was not that it was another fundamental particle but rather a combination of proton and electron in a "neutral doublet." Back in the Cavendish, he often talked about the idea of the neutron to his chief assistant, James Chadwick, as they sat in a darkened laboratory waiting to count scintillations from a zinc sulphide screen, which was Rutherford's preferred method of detecting charged subatomic particles. Over the years, Rutherford seemed to lose interest in the notion of a neutron, but not so Chadwick who surreptitiously pursued a number of false trails, trying to find evidence for its existence. This type of speculative research was not to Rutherford's taste, and the shy Chadwick would not have disclosed his unsuccessful endeavors to his chief. Two of Chadwick's research students discovered an anomalous result, which he thought was likely to be due to the presence of neutrons, but the crucial breakthrough came in early 1932, when he read a paper from the Joliot-Curies, who worked in Paris. They had interpreted some surprising observations of their own in terms of gamma radiation of unprecedentedly high energy. As soon as he read their paper, Chadwick rushed in to tell Rutherford that the Joliot-Curies had in fact missed the neutron.

At once, Chadwick set about designing his own experiment and within weeks, using apparatus that could have been built by a talented schoolboy, he had conclusive proof of the existence of the neutron. The discovery was announced in a concise letter to *Nature*, and the birth of the neutron transformed what had become a rather muddled subject into the more coherent field of nuclear physics. A few months later, Cockcroft and Walton (also at the Cavendish) became the first to disintegrate an atomic nucleus by bombarding it with artificially accelerated particles. The following year, Chadwick and Cockcroft were among the leading

Sir James Chadwick, as Master of Gonville and Caius College (1948–1958). (Andrew Brown, *The Neutron and the Bomb* (Oxford University Press, 1997).)

participants at the Seventh Solvay Conference on the atomic nucleus, which marked the last time the pioneers of radioactivity like Rutherford and Marie Curie met with the new giants of nuclear physics such as Fermi and Heisenberg: there was a lone American at the conference — a young professor from U.C. Berkeley, Ernest Lawrence, who talked about his new invention, the cyclotron.

The Solvay conference was held in Brussels in October 1933, by which time the lives of hundreds of European scientists had already been disrupted by the ascension to power of Hitler and the German Nazis. Scientists in England were generous in offering their colleagues refuge and the opportunity to continue to work; Rutherford himself had been appointed the first President of the Academic Assistance Council. Two early arrivals at Manchester University were the theoreticians, Hans Bethe and Rudolf Peierls: they would be followed by dozens of others, some like Bethe who soon moved on to the USA, others like Peierls who became British citizens and lived in England for the rest of their lives. One of the last scientists to reach England was Otto Frisch, who had been half-heartedly seeking sanctuary for some time (he was from a well-known Jewish family in Vienna); he was finally invited to Birmingham University by Mark Oliphant in the summer of 1939.

Frisch, then aged 35 years, had suddenly achieved scientific fame in January 1939, when he wrote a letter to *Nature* describing the new phenomenon of *nuclear fission* — Frisch's was the second name on the letter, the first being that of his aunt, Lise Meitner. Where Rutherford and his school had split the atom to the extent of chipping or shaving the nucleus to leave a closely related atom, Meitner and Frisch described a new type of nuclear reaction — the disintegration of the heaviest atom then known, uranium, into much lighter atoms such as barium. The effect was achieved by bombarding uranium with neutrons, and resulted in a uranium nucleus dividing into two nuclei of roughly equal size (as a cell divides in the biological fission process): in the process mass is converted into energy, as described by Einstein's famous equation.

Apart from the release of energy, the fission process also generated more neutrons so that the question of a possible chain reaction was raised, but this seemed not to be a realistic proposition because Bohr and Wheeler had theorized that the fission process was triggered by slow neutrons in uranium-235, a rare isotope that constitutes less than 1% of natural uranium. Rudolf Peierls, by then at Birmingham University, was still concerned about the possibility of a chain reaction leading to a very powerful weapon. He consulted Frisch, who, in the spring of 1940, posed the following question to him: "What would happen if somebody gave you the pure isotope uranium-235?" The two men quickly calculated that the critical mass for a chain reaction would now be in the order of pounds, and not many tons as previously assumed. The second critical assumption they made was that the chain reaction would require fast, not slow, neutrons. As "enemy aliens," Frisch and Peierls were not allowed to take part on the top secret radar research taking place in Birmingham, and left to their own devices they developed their ideas on a super-bomb in a two part memorandum (typed by Peierls).

The Frisch–Peierls Memorandum must rank as one of the most historic documents of the twentieth century. It was written at a time when the war was still confined to Europe — the Soviets, the Japanese and the Americans were all interested bystanders — and the English Channel seemed to become narrower each week to the British and the Germans. Not only is the Frisch–Peierls Memorandum remarkable as the first technical exposition of an atomic weapon, the two authors also considered the military and ethical consequences of such a super-bomb in a simple but profound way. Among the points they made were the following:

1. As a weapon, the super-bomb would be practically irresistible. There is no material or structure that could be expected to resist the force of the explosion.

2. Owing to the spreading of radioactive substances with the wind, the bomb could probably not be used without killing large numbers of civilians, and this may make it unsuitable as a weapon for use by this country.
3. We have no information that the same idea has also occurred to other scientists but since all the theoretical data bearing on this problem are published, it is quite conceivable that Germany is, in fact, developing this weapon.

Frisch and Peierls handed their document to Oliphant who scrutinized it before sending it to London. A small committee, originally consisting of five men who were all alumni of Rutherford's Cavendish Lab., was appointed to consider the uranium bomb. For a short period, security restrictions meant that the two authors of the memorandum were excluded from any further discussions, until Peierls wrote a letter pointing out that he and Frisch had immersed themselves in this subject for some months, and had already made crucial discoveries overlooked by others. The point was taken and they were both fully consulted by what had become known as the M.A.U.D. Committee, and became members of the Technical Sub-Committee under Chadwick. This small group were faced with the immense challenge of gathering experimental data to support the ideas outlined in the Frisch–Peierls Memorandum.

The scientific problems were legion and difficult: could uranium-235 be successfully separated from the commoner 238 isotope, and what was the true critical mass that would be needed to produce a weapon? While the M.A.U.D. Committee met during 1940, Churchill became Prime Minister, France fell to the Nazis, the RAF was stretched to its limits in the Battle of Britain. The experimental work was carried out in small university departments, much of it in Liverpool where Chadwick was now Professor of Physics. He had one telephone, limited petrol rations, and a department that sustained frequent bomb damage. In the blitz of May 1941, Liverpool was bombed seven nights in a row, there were three thousand civilian casualties and seventy thousand were left homeless.

In July 1941, Chadwick wrote the Maud Report which chronicled the scientists' collective journey from initial skepticism to a firm belief that "it will be possible to make an effective uranium bomb which, containing some 25 lb of active material, would be equivalent as regards destructive effect to 1,800 tons of TNT." Chadwick also realized that the manufacture of this weapon would require a gigantic industrial enterprise: only one member of the M.A.U.D. Committee, Patrick Blackett, thought this was beyond the means of war-ravaged Great Britain.

General Groves and Sir James Chadwick in Washington, D.C. (Andrew Brown, *The Neutron and the Bomb* (Oxford University Press, 1997).)

A copy of the Maud Report was sent to the Uranium Committee in the U.S.A., but seemed to provoke no interest. Later that summer, Mark Oliphant flew to the U.S.A., primarily to review progress in radar research. As a member of the M.A.U.D. Committee, he made inquiries about the fate of their report. He was "amazed and distressed" to learn that the elderly chairman of the Uranium Committee, Lyman Briggs, had locked it away in his safe for security and had not shown it to anyone. Oliphant made a detour from Washington to Berkeley to see Lawrence and convinced him that the Maud work was crucially important. Lawrence was easily convinced and lit a fire under the American scientific establishment that within a year led to the Manhattan Project.

A Footnote on Hiroshima and Atomic Morality: Conant, Niebuhr, and an "Emotional" Clergyman, 1945–46

James G. Hershberg
Department of History, George Washington University

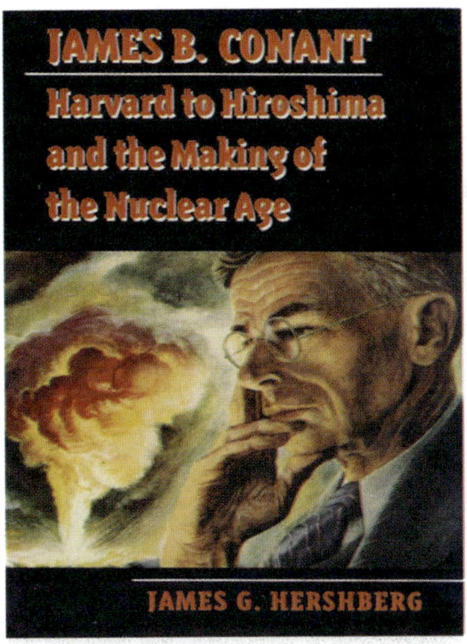

One of the most prominent, if private, debates about the morality of using the atomic bomb occurred in an exchange of letters in March 1946 between the era's most prominent educator and most prominent theologian: James B. Conant — President of Harvard University and, as an official of the Office of Scientific Research and Development and then as a member of the Interim Committee, a key figure in the decisions to build and use the atomic bomb in World War II — and Reinhold Niebuhr, then a professor at Columbia University.

Conant, who as a member of the Interim Committee had endorsed the bomb's use (and was recorded in the minutes of its May 31 meeting as suggesting the criteria of using the weapon on "a vital war plant employing a large number of workers and closely surrounded by workers' houses"), was upset on the morning of March 6, 1946, by a report on the front page of *The New York Times* that listed Niebuhr among the signatories of a draft declaration by the Federal Council of Churches calling the use of the bomb "morally indefensible." An admirer of Niebuhr who had vainly tried to lure him to Cambridge, Conant had been especially disturbed to read of the theologian's position because, like many liberals who had supported intervention in World War II, he had often cited Niebuhr's writings on the permissibility of using immoral means to achieve moral ends to defend U.S. participation in the conflict against Nazi Germany.

In an unusually impassioned (for him) letter, Conant sharply defended the atomic bombings as being no more immoral than strategic incendiary bombings of cities (noting that "I was as deeply involved in the one method of destruction as the other," so at least on that point he could be impartial) or many other violent actions committed by the United States in the process of winning the war. Moreover, with fears of an eventual conflict with Russia increasing, Conant worried that disavowing the bomb's use on Japan implicitly meant foreswearing such weapons for the future, and unilaterally disarming — "a logical and defensible position, but to my mind unrealistic."

While Niebuhr robustly defended the desirability of acknowledging some "expression of guilt" regarding the bomb's use — "I thought it important from the Christian standpoint to admit the moral ambiguity of all righteous people in history, who are, despite the good they do, involved in antecedent and in marginal guilt" — Conant's protests clearly had some impact, since the theologian subsequently acted to tone down the Council's statement on the bomb.

While some have wondered whether Conant's letter represented, as he frankly acknowledged to Niebuhr, "a highly personal reaction by one who has a guilty conscience," the exchange can also be seen as a highly symbolic conversation reflecting the efforts of leading American liberals, who had favored the war (and would likewise support the Cold War) and opposed isolationism, to grapple with the problems of ends vs. means in the atomic age, and the integration of nuclear weapons into America's moral as well as military arsenal as it headed into the postwar era of global leadership and possibly renewed conflict. The exchange has been recounted in several works, including Richard Fox's biography of Niebuhr and my own work on Conant.

What I present here for the first time, however, is an antecedent exchange between Conant and a far less prominent religious figure, which sheds light, I think, not only on the intensity of his reaction to Niebuhr's statement but on his own rationale for using the atomic bomb in the first place. In a handwritten postscript to his March 6 letter to Niebuhr, Conant alluded to having received a few months earlier "a very emotional letter from a clergyman denouncing me for my part in the atomic bomb development." In response, Conant had urged him to read Niebuhr's *Children of Light and Children of Darkness*, which stoutly defended the morality of going to war to defend civilized values. Now, Conant told Niebuhr, "I can't reconcile this book with your signature on the document in question."

President of Harvard University James B. Conant after World War II. (James G. Hershberg, *James B. Conant: Harvard to Hiroshima and the Making of the Nuclear Age* (Stanford University Press, 1993).)

Though Conant had alluded only vaguely to the clergyman at issue, I had discovered the final letter in the exchange to which he referred to Niebuhr in time to include it in my book. In it, on December 13, 1945, the Rev. Bradford Young of Grace Church in Manchester, New Hampshire thanked Conant for his "patient answer to my somewhat excited letter." Rev. Young acknowledged that he "largely followed" the reasoning in Niebuhr's *Children of Darkness and Children of Light*, and that the "A-bombing was no worse in its effects than the obliteration bombing." However, clearly alluding to previous correspondence, Rev. Young reacted negatively to what he described as Conant's argument that it had been necessary to drop the bomb on Japanese cities in order to alert world public opinion to the danger of future atomic war so that measures could be taken to put the weapon under international control. Considering the human cost to the inhabitants of Hiroshima and Nagasaki, Young had written, this was a calculation that "only God" could make. "What bothered me," he added, was to see you preparing and participating in such a Godlike decision with *apparently* no sense of presumption, no fear and trembling, no feeling of tragic involvement in a horrible deed." Unfortunately, at the time I was finishing *James B.*

Conant (the early '90s), I had been unable to locate Young's original "somewhat excited letter," nor Conant's response.

However, during a later visit to Cambridge, I returned to the Harvard Archives — where, under the University's stringent 50-year secrecy rule, the papers for 1945–46 did not open up completely until the summer of 1996 — and found the rest of the exchange. The exchange had been prompted by a presentation Conant gave on December 3, 1945, to the Harvard Club of Manchester, New Hampshire. One member of the audience was Rev. Young, an Episcopalian minister (and Harvard graduate) active in social causes and described by one family member as a "Norman Thomas–style socialist" (though not a formal party member) and a "very serious pacifist."

During his talk, Conant showed photographs of the destruction in Hiroshima and Nagasaki. "As the pictures told their terrible story," Young wrote Conant the next day, "I felt that all of us there were war criminals reenacting a scene strangely like that when Goering *et al.* were compelled to view the pictures of the barbarities of Buchenwald." Like the Germans, Young went on, Americans had "closed their eyes to what we had done or what was done in our name," "not daring to speak or even think of the scores of thousands of our brother humans who perished by our hand. Or we said, as cruel people have often justified their savagery, it shortened the war." If there were no good reason why "the bomb could not have been demonstrated to the Japanese as persuasively yet as harmlessly" as it had been to Conant in New Mexico, then "the crime you helped us all commit was of the same stupendous order of the bomb." Perhaps Conant's well-known calm demeanor further infuriated Young, for he added: "If you have a conscience about that crime, you concealed it wonderfully well. If you have none, it's monstrous." Though described by a family member as having a "rational clerical" rather than an emotional manner, a clearly outraged Young signed off "With hopes for your repentance and all the world's."

In his response, dated December 7, Conant did indeed recommend consulting Niebuhr's text, stressed as he did elsewhere the inextricable linkage between the acceptance of the ethics of strategic bombing of cities and the use of the atom bomb, and described his own rationale for using the weapon as two-fold: "first, because it was a valuable supplement to the strategic bombing then in progress and which I hoped would end the war without an invasion; and second, because I felt certain that unless this bomb was demonstrated in combat there was very little chance of arousing public opinion to a point where they would take sufficiently drastic action to control it in the future."

That Conant gave equal prominence to this postwar rationale as well as to wartime military imperatives (which would become the orthodox or traditional defense for the bomb's use) is significant, for it reflected that strong sense of fear animating him as well as many of the scientists who worked on the Manhattan Project (including Oppenheimer) of a postwar nuclear arms race and eventual nuclear World War III. The simple argument that it was needed to defeat Japan and avoid an invasion was not enough, in other words, if the consequence would be to trigger a postwar nuclear arms race among the former Allies, i.e., between the U.S. and the Soviet Union.

Vannevar Bush, James Conant, and General Leslie Groves inspecting Hanford during the Manhattan Project. (James G. Hershberg, *James B. Conant: Harvard to Hiroshima and the Making of the Nuclear Age* (Stanford University Press, 1993).)

Some atomic scientists, particularly among those at the Met. Lab in Chicago, believed that dropping the bomb without warning on Japanese cities would be most likely to cause such a disastrous competition — and they propounded their views in the summer of 1945 in the Franck Report, which was submitted to the Interim Committee but never reached Truman's Desk. (The text has been published in various places, most conveniently as an appendix to the most recent paperback edition of Martin J. Sherwin's *A World Destroyed*.) However, Oppenheimer and other leading atomic scientists (including Fermi, Lawrence, and Arthur Compton) dissented, finding no plausible alternative to use of the bomb on Japanese cities without prior warning — and so reported as the Scientific Panel to the Interim Committee. And, as physicists interviewed in the documentary *The Day After Trinity* recall, Oppenheimer explicitly argued to scientists at Los Alamos that the bomb's use in this fashion represented the *best* chance to convince the world to accept international control after the war. Conant, too, I believe, came to this position, although the minutes of the Interim Committee do not record him as specifically making this argument (not entirely surprisingly, since even if he did make such a case the only known discussion over whether or not to use the bomb at all came during an informal lunchtime conversation). While at least one commentator has questioned whether Conant really held this position, this new letter to Young provides an additional, albeit *post hoc*, piece of evidence to suggest that he genuinely did.

However, it remains unclear whether in his heart of hearts Conant sincerely believed that using the bomb on Hiroshima and Nagasaki was the best course to follow to head off a nuclear arms race leading to a nuclear World War III — as early as May 1944 he had privately feared that humanity's only alternatives were a "race between nations and in the next war destruction of civilization, or a scheme to remove atomic energy from the field of conflict" — or whether he was rationalizing to himself his participation in, and support for, a decision and action he now viewed as inevitable and politically impossible to oppose; or, for that matter, whether Conant really knew himself.

A Los Alamos Beginning[24]

Kai Bird
Fellow at the Woodrow Wilson International Center for Scholars

Martin Sherwin
Professor of History at Tufts University

Robert Oppenheimer had always wanted to combine his passion for physics with his fierce attraction to the desert high country of New Mexico. And now he had his chance. On November 16, 1942 he and Edwin McMillan, another Berkeley physicist, accompanied Army Major John H. Dudley to Jemez Springs, a deep canyon forty miles northwest of Santa Fe. After inspecting dozens of potential sites across the American southwest, Dudley had finally settled on Jemez Springs as a suitable home for the proposed new weapons laboratory. Oppenheimer remembered it from his horseback trips as a "lovely spot and in every way satisfactory."[25]

But when the three men arrived at Jemez Springs, he and McMillan began arguing with Dudley that the snake of land at the bottom of the canyon was too narrow and confined for the little town they envisioned building. "He [Oppenheimer] took one look at Jemez Springs," Dudley recalled, "and proceeded to change the criteria." Where once hills were necessary, now Oppie said, "No hills." Dudley had thought a good paved access road was an absolute must — but now Oppenheimer argued that all that was required was a road good enough to haul up two heavy howitzers. Dudley knew that could be done on a dirt track, "So the problem of access was practically wiped out."[26]

[24] An excerpt from Kai Bird and Martin Sherwin's forthcoming biography of J. Robert Oppenheimer to be published by Alfred A. Knopf.

[25] J. Robert Oppenheimer, *Robert Oppenheimer: Letters and Recollections (Stanford Nuclear Age)*, eds. Alice K. Smith and Charles Weiner (Stanford, CA: Stanford University Press, 1995), p. 236.

[26] Gen. John Dudley, public lecture, "Reminiscences of Los Alamos, 1943–1945," 13 Mar. 1975.

Oppenheimer complained that it had no view of the magnificent mountain scenery — and that the site's steep canyons would make it nearly impossible to fence. "We were arguing about this when General Groves showed up," recalled McMillan. Groves took one look at the site and said, "This will never do." Dudley thought Groves was just unhappy that they had not settled on a site: "So he asked Oppenheimer if there was something else around that had prospects. Oppie proposed Los Alamos as though it was a brand new idea." (Actually, Oppenheimer had already visited Los Alamos the previous July, accompanied by Groves's immediate predecessor, Col. James C. Marshall, an Army Corps of Engineers officer. For Oppenheimer, Los Alamos had always been a prime site.)[27]

"If you go on up the canyon," Oppenheimer told him, "you come out on top of the mesa and there's a boys' school there which might be a usable site." Reluctantly, the men piled back into their cars and drove northwest about thirty miles across a lava mesa called the Pajarito (Little Bird) Plateau. It was already late afternoon when they pulled up to the Los Alamos Ranch School. Through the haze of a light, drizzly snowfall Oppenheimer, Groves and McMillan saw a group of schoolboys out on a playing field running around in shorts. McMillan remarked that the headmaster must really believe in "hardening up the youth." The ranch school's 800 acre grounds included the "Big House," the school's main building, Fuller Lodge, a beautiful manor house built in 1928 from 800 huge Ponderosa logs, a rustic dormitory and a few other smaller buildings. Behind the lodge there was a pond that the boys used for ice skating in the winter and canoeing during the summer. The school stood at an elevation of 7,200 feet, just about at timberline. To the west, the snowcapped Jemez Mountains rose to roughly 11,000 feet. From the spacious porch of Fuller Lodge one could look forty miles east across the Rio Grande Valley to Oppenheimer's beloved Sangre de Cristo mountain range, rising to a height of 13,000 feet.

[27] Marjorie Bell Chambers and Linda K. Aldrich, *Los Alamos, New Mexico, A Survey to 1949 (The Los Alamos Story, Monograph 1)* (Los Alamos: Los Alamos Historical Society, 1999), p. 27. Chambers and Aldrich report that Percival C. Keith, a member of the Planning Board of the wartime Office of Scientific Research and Development, had also suggested Los Alamos to Col. Marshall. Keith had a son who had attended the Los Alamos Ranch School, which explains why he was already familiar with the site.

By one account, as Groves surveyed the scene, he suddenly announced, "This is the place." And yet, Dudley later insisted that Groves was not too pleased by the site, but being pressed for time, he left the men to discuss the issue. Later that night Dudley reported back to him the general tenor of his conversation with Oppenheimer. "I told him," Dudley said, "that the criteria would be continuously modified until only Los Alamos fit the criteria. So I said, 'Let's pick Los Alamos and get on with our other work.'" Groves agreed.[28]

Within two days the army initiated the paper work to buy the school and four days later, after a quick trip to Washington, D.C. and back, Oppenheimer returned with McMillan and Ernest Lawrence to inspect what had been designated "Site Y." Wearing cowboy boots, Oppenheimer took Lawrence on a tour of the school buildings. For security purposes they had introduced themselves under assumed names. But a Los Alamos student, Sterling Colgate, recognized the scientists. "Suddenly we knew the war had arrived here," Colgate recalled. "These two characters showed up, Mr. Smith and Mr. Jones, one wearing a porkpie hat and the other a normal hat, and these two guys went around as if they owned the place." A senior in high school, Colgate had studied physics and knew Oppenheimer was a leading theoretical physicist and recognized Lawrence as the great builder of cyclotrons. Even as an armada of bulldozers and construction crews invaded the school grounds in the coming weeks, Colgate realized that the army was putting "megabucks" into "what seemed to us the worst place in the world to have a laboratory because there was no railroad and no water or any of those things that you normally need for a laboratory."[29] The nearest real town, Santa Fe, was forty miles distant.

---

[28] John H. Manley, "A New Laboratory Is Born," unpublished manuscript, p. 13; Edwin McMillan, "Early Days of Los Alamos," manuscript, p. 7; John H. Dudley, public lecture, "Ranch School to Secret City," 13 Mar. 1975, reproduced in: Lawrence Badash, Joseph O. Hirschfelder, and Herbert P. Broida, eds., *Reminiscences of Los Alamos, 1943–45* (Dordrecht, Holland: D. Reidel Publishing Company, 1980).

[29] Sterling Colgate interview (John Else), 12 Nov. 1979, 2–3. The initial construction contract was for $300,000 — but within a year $7.5 million had been spent. (John Dudley, public lecture, 1943–1945, 13 Mar. 1975). That Oppenheimer was wearing cowboy boots is described in: Peggy Pond Church, *The House at Otowi Bridge: The Story of Edith Warner and Los Alamos* (Albuquerque, NM: University of New Mexico Press, 1959), p. 84.

Oppenheimer, of course, knew Los Alamos well. His ranch was a mere forty miles across the mesa. He and his brother had explored the Jemez Mountains on horseback over many summers. He knew exactly how Spartan and isolated Los Alamos was from the rest of the world — and he wanted it that way. McMillan thought he may have had Los Alamos in mind all the time. "I think Oppenheimer privately wanted that site anyhow," McMillan recalled, "but he didn't put it forward until this occasion."[30]

Oppenheimer got what he wanted — a spectacular view of the Sangre de Cristo Mountains — and General Groves got a site so isolated that there was only a winding gravel road and one phone line into the place. Over the next three months construction crews threw up cheap barracks with shingled or tin roofs. Similar buildings were thrown up to serve as crude chemistry and physics laboratories. Everything was painted Army green.

Oppenheimer seemed unaware of the utter chaos that had descended on Los Alamos. Focused on recruiting the scientists he needed for the project, he had no time for the mundane administrative tasks associated with building what was essentially a small town. John Manley, a seasoned experimental physicist whom Oppie had tapped as one of his assistants, had serious qualms about the site. Manley had just come from Chicago where on December 2, 1942 the Italian émigré physicist Enrico Fermi conducted the world's first controlled nuclear chain reaction. Chicago was a big city with an eminent university, libraries, and a large pool of experienced machinists, glass-blowers, engineers and other technicians. Los Alamos had nothing.

"What we were trying to do," wrote Manley, "was build a new laboratory in the wilds of New Mexico with no initial equipment except the library of Horatio Alger books or whatever it was that those boys in the Ranch School read, and the pack equipment that they used to going horseback riding, none of which helped us very much in getting neutron-producing accelerators." Manley thought that if Oppenheimer had been an experimental physicist he would have understood that "experimental

[30] Edwin McMillan, "Early Days of Los Alamos," unpublished manuscript, p. 8.

physics is really 90 percent plumbing" — and he never would have agreed to building a laboratory in such a setting.[31]

The logistics were horrendously complicated. Oppenheimer and the initial group of scientists planned to arrive at Los Alamos by mid-March, 1943. By then Oppenheimer assured Hans Bethe — whom he was trying to recruit in December — that he was building a viable community run by a city engineer. There would be bachelor quarters and homes for families with one, two and three bedrooms. These furnished quarters would all come with electricity — but for security reasons there would be no phones. The kitchens would be equipped with wood-fired stoves and wood-fired hot water heaters. There would be fireplaces and an electric icebox. Servants would be available on occasion for any heavy housework. There would be a school for young children, a library, a laundry, a hospital and garbage collection. An army Post Exchange would serve as the community's grocery store and mail order house. "The Exchange," Oppie explained to Bethe, "will be able to order for us what they do not carry." A recreation officer would arrange for regular movies and hiking trips in the near-by mountains. And Oppie promised there would be a cantina for beer, cokes and light lunches, a regular mess hall for unmarried people and a "fancy" café where married couples could eat out in the evening.[32]

For the laboratories, they ordered the shipment of two Van de Graaff generators from Michigan, a cyclotron from Harvard and a Cockcroft–Walton machine from the University of Illinois. All three machines were essential pieces of equipment. The Van de Graff generators would allow them to run basic physics measurements. And the Cockcroft–Walton machine, the first particle accelerator, would allow them to conduct experiments in which various elements could be artificially transmuted into other elements. (This was the machine used in the first experiments that figuratively "split the atom" by transforming lithium elements into helium elements.)

---

[31] John H. Manley, "A New Laboratory Is Born," unpublished manuscript, p. 18.
[32] Smith and Weiner, eds., pp. 244–245. JRO to Hans and Rose Bethe, 28 Dec. 1942.

The construction of Los Alamos, the recruitment of scientists and the assembling of all the equipment necessary for the world's first nuclear weapons laboratory required a meticulous and patient administrator. In early 1943, Oppenheimer was neither. He had never supervised any organization larger than his graduate seminars. Neither, thought his peers, did he seem temperamentally suited for the job. "He was something of an eccentric — almost a professional eccentric when I knew him before 1940," recalled Robert Wilson, a young experimental physicist who was then studying under Ernest Lawrence. "He just wasn't the kind of person that you would think would be an administrator."[33]

Even some of his closest friends were skeptical that their Oppie could do this. One evening Oppenheimer was sitting around with Bob and Charlotte Serber and his new young secretary, Priscilla Greene, discussing the project. He and Serber were tossing out names of people they could recruit when Charlotte suddenly interrupted and said, "You people think you're going to do this? You?"[34]

John Manley had serious misgivings about serving as Oppie's deputy. "I had only briefly met him," Manley recalled. "I had given a colloquium in Berkeley a year or two before and I was somewhat frightened of his evident erudition and his lack of interest in mundane affairs." Manley quickly became worried about the laboratory's organization. "I bugged Oppie for I don't know how many months about an organization chart — who was going to be responsible for this and who was going to be responsible for that." Oppenheimer ignored his pleas until finally one day in March 1943 Manley climbed to the top floor of LeConte Hall and pushed open the door of Oppenheimer's office. Ed Condon was sitting there, a good sign, thought Manley, because Condon had some experience from his work at the Bureau of Standards. When Oppenheimer glanced up to see Manley standing there he knew exactly what he wanted. Grabbing a piece of paper, he threw it down on his desk and said, "Here's your damned organization chart."[35] Oppenheimer envisioned four broad

[33] Robert R. Wilson, interview with Owen Gingrich, 23 Apr. 1982, 3.

[34] Priscilla Duffield, interview by Alice Smith, 2 Jan. 1976, 9.

[35] John H. Manley, interview by Sherwin 9 Jan. 1985, 23; John H. Manley, "A New Laboratory Is Born," unpublished manuscript, p. 21.

divisions within the laboratory: experimental physics, theoretical physics, chemistry and metallurgy, and finally, ordnance. Group leaders within each of these divisions would report to the division chiefs — and the division chiefs would report to Oppenheimer. It was a beginning.

In early 1943, Oppenheimer sent Robert Wilson, 28, to Harvard with the responsibility to arrange for the safe shipment of Harvard's cyclotron to Los Alamos. On March 4, 1943 Wilson arrived in Los Alamos to inspect the building that would house the cyclotron. He found utter chaos; there seemed to be no schedule, no planning and no line of responsibility. Wilson complained about the situation to Manley and the two men agreed they should confront Oppenheimer. "So we went to see him in Berkeley," Wilson recalled. "Well, it seemed in those days that much that Oppenheimer did was done at a party; at least that's where we cornered him. He'd been doing his usual magic with the martinis — just the right amount of martini and a proper amount of gin and made very ceremoniously, and whether he'd had too many or it was Manley and I, I don't know, but we got into a terrible argument. We just went after Oppie unrelentingly with all these practical problems that we were trying to make him understand, and after a certain stage, Oppenheimer became extremely angry. He began to use vile language, asking us why we were telling him of these insignificant problems, that it was none of our business, and so on. Both of us were scared to death. We were frightened because, if this was the leader and, if the leader was going to have a tantrum to resolve a problem, then how was anything going to be sorted? So we withdrew, John and I, and discussed some more and decided that we would take more initiative and not look for so much leadership from Oppy."[36]

A Quaker by ancestry, Wilson was a pacifist when the European war erupted: "So it was quite a change for me to find in fact that I would be working on this horrible project." But like everyone else he knew at Los Alamos, Wilson feared the Nazis could win the war with an atomic weapon. And while privately he still hoped that they might someday

---

[36] Peter Goodchild, *J. Robert Oppenheimer: Shatterer of Worlds* (Warsaw: Fromm International, 1985), p. 72.

prove that an atomic bomb was not possible, he was eager to build it if it could be built.[37] Hardworking and serious minded by temperament, Wilson initially found himself annoyed by Oppenheimer's arrogant demeanor. "I sort of disliked him," he later said. "He was such a smart-aleck and didn't suffer fools gladly. And maybe I was one of the fools he hadn't suffered."

And yet, within a short time Wilson found himself changing his assessment of Oppenheimer. He was surprised to see that his eccentric boss had somehow metamorphosed into a consummate administrator. "He had style and he had class," Wilson said. "He was a very clever man. And whatever we felt about his deficiencies, in a few months, he had corrected those deficiencies, and obviously knew a lot more than we did about administrative procedures. Whatever our qualms were, why, they were soon allayed."[38] By the summer of 1943 Wilson noticed that "when I was with him, I was a larger person... I became very much of an Oppenheimer person and just idolized him... I changed around completely."[39]

[37] Mary Palevsky, *Atomic Fragments: A Daughter's Questions* (Berkeley, CA: University of California Press, 2000), pp. 128–129.

[38] Robert R. Wilson, interview with Owen Gingrich, 23 Apr. 1982, 4.

[39] Palevsky, 134–135; Robert R. Wilson interview with Owen Gingrich, 23 Apr. 1982, 4 (Sherwin collection).

Chapter 4: The Military and Science in the Crucible of War

General Leslie R. Groves and the Scientists

Robert S. Norris
Natural Resources Defense Council

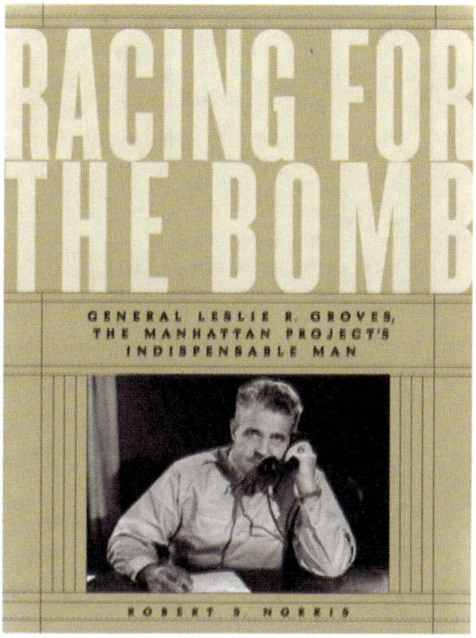

My short paper discusses certain aspects of General Leslie R. Groves' relationships with the scientists. One caricature of Groves (of many) is that he and the scientists were in constant disagreement, always at loggerheads, with widespread differences over matters of discipline, censorship, and security. It was a clash of cultures between the military and science, never resolved.

The actual relationships are quite different, more complex and nuanced. They range from outright hostility, in the case of Leo Szilard, to ones of deep respect and appreciation as in the cases of Robert Oppenheimer, Richard Tolman, and many others. For Groves, scientists came in many varieties. There were the troublemakers like Szilard, and later James Franck; the incompetents like the administratively-challenged Harold Urey; the grand planners and organizers like Ernest Lawrence; and the icons like Niels Bohr, who commanded respect for their sheer brilliance. While most of the scientists were American, there were also British, French, Italian, Hungarian and other foreign scientists who caused him concern. The criterion for a scientist to be in Groves' good graces was a willingness to get the bomb built as quickly as possible. Anything else was a distraction, from speculating on the mysteries of the atom to meddling in political matters, where he thought they had no business.

The first contact between Groves and the scientists came less than three weeks after he was chosen to head the Manhattan Project. At the beginning of October 1942, Groves set out to tour the university laboratories in Chicago, Berkeley, and New York. On October 5, 1942, Groves was at the Metallurgical Laboratory at the University of Chicago for an initial meeting with the scientists. Much would happen that day that would set the tone for the relationship between Groves and the scientists. Impressions would take root (on both sides) that would be long lasting. He found some of them arrogant and others naïve, believing they could produce the bomb material in their tiny laboratory, apparently unable to grasp the huge engineering-industrial effort that would obviously be required.

Groves was astounded that no one could answer the practical question of how much fissionable material — either highly enriched uranium or plutonium — would be needed for a bomb. Expecting a reply of within twenty-five or fifty percent he was astounded when they told him that they could only estimate the amount within a factor of ten. In other words, if it was determined that 100 pounds of uranium might be needed for the bomb, the correct amount could be anywhere from 10 pounds to 1,000 pounds, not the best guide for an engineer who had to build the factories to produce it.

That meeting at Chicago is usually presented as Exhibit A in fashioning the caricature of Groves. For some the impression of the General as an uneducated military martinet was sealed. They were horrified and angry that the project on which they had been working was being taken over by the army, and worse yet, was going to be run by this particular general.

Groves for his part was not out to win a popularity contest. In fact, he was already a bureaucratic warrior of the first rank, having overseen all Army construction during the mobilization period and early months of the War. Having a tough and intimidating reputation is just a wise strategy for getting your way and getting things done. Charm was not among his personality attributes. Underneath that hard exterior was an even harder inner core, a "will of steel" a West Point classmate called it. He was very judgmental about people, and sometimes he let them know it. His standards were difficult for most people to achieve. Groves readily accepted people if he thought them intelligent and competent, and pretty much dismissed those who were not. He liked people like himself: organizers, people who were decisive and not afraid to make difficult choices and, at times if necessary, even take risks. He sized people up quickly and decided that they either could do the job or they couldn't. There were few second chances.

Much has been made of Groves' concern with secrecy and his widespread use of compartmentalization. While Groves did not invent it, he implemented it on a scale not previously seen. As I argue in the book, for Groves compartmentalization was not only a method to maintain secrecy, it was also the source of his power. All of the boxes in the wiring diagram led to him. He, and he alone, knew everything and because of that was able to shape the substance and the pace of events. Some of the scientists claimed that compartmentalization hindered the free interchange of ideas that is essential to scientific discovery and progress, and thus slowed development of the bomb. It is true that any communication between the Manhattan sites had to go through Groves' channels and any trips were strictly regulated. Scientists could not just travel between Chicago and Los Alamos, for example, or from Oak Ridge to Hanford, without permission from Groves. But when there was a good reason for a trip, he gave it. When he thought that a meeting of individuals or groups could speed up the project, he was flexible. Otherwise, as he said, "everyone should stick to their knitting."

General Leslie R. Groves.

When we turn to Groves' relationships with individual scientists we see that his experience with Szilard was the exception. With most of the others with whom he worked, Groves got on well. In Washington, his closest contacts were with Vannevar Bush, James Conant, and Richard Tolman.

After some difficulties at their first meeting Bush and Groves worked well together. They were both engineers, practical men who were focused on the job at hand. They saw eye to eye on most things. Bush fully supported Groves's imposition of secrecy and security and was a man of action when it came to trying to halt the German bomb. He was an early advocate of aerial bombing and later fully supported the Alsos mission to gather information about the German bomb program, supplying resources and personnel. Bush saw his role as keeping President Roosevelt informed on the status of the project and giving him his best advice. Equally important, as head of the OSRD Bush helped supply Groves with any scientist in America that he might need.

At the outset, Groves knew very little about nuclear physics and the complex technologies involved. But whatever the topic, Groves always did his homework. It was not his style to leave anything to chance if it could be avoided. He knew that in many cases he was going to have to rely on the judgment and advice of others. Therefore, he would have to learn at least the basics and find scientists whose opinions he could trust.

On June 11, 1943, Groves formalized an arrangement with Conant and Tolman, asking them to serve as his scientific advisers. In response to his probing questions, Conant and Tolman provided Groves with detailed and comprehensive primers on all aspects of nuclear technology. Armed with this kind of advice, Groves was able to make several fundamental decisions on technical matters that affected the basic course of the project.

Tolman played a significant role in the Manhattan Project, one that has not been fully appreciated. Groves relied upon him extensively. Tolman was a well-respected scientist in his own right and knew all of the major personalities in the field. He assisted in the constant recruitment that sought out the best scientists. He traveled to the major sites, smoothed the friction between the scientists and the military, and drafted countless memos and policy papers for Groves.

Conant and Tolman were not the only sources of counsel and opinion for Groves. The three of them were able to draw upon some of the best scientists in the world. Eight on the project had already won a Nobel Prize, and more than a dozen others would do so after the war. As Groves said, "...vital decisions were reached only after the most careful consideration and discussion with the men I thought were able to offer the soundest advice. Generally, for this operation, they were Oppenheimer, [John] von Neumann, [William] Penney, [William] Parsons and [Norman] Ramsey." But after all was said and done, when a big decision had to be made it was Groves's call. Often it fell to him to make a choice between what appeared to be equal alternatives, each backed by sound scientific advice.

Groves's most important relationship with a scientist was the one with Oppenheimer. They met on Groves's initial tour of the laboratories and from the outset they thought along similar lines, about where to establish a laboratory, about how to organize it, and how to run it.

Often we have chosen to see these two as the "odd couple," wondering how two people so different could get along: the portly military engineer and the gaunt theoretical physicist, who read Sanskrit for relaxation. While they might have been different in some ways, they were much alike in others.

That Oppenheimer and Groves should have worked so well together is really no mystery. Groves was especially perceptive in sizing people up, usually within a matter of minutes and knowing whether they could do a job or not. In Oppenheimer he saw an "overweening ambition" that drove him. He understood that Oppenheimer was frustrated and disappointed; that his contributions to theoretical physics had not brought him the recognition that he believed he deserved and craved. This project could be his route to immortality. Part of Groves' genius was to entwine other people's ambitions with his own. Groves and Oppenheimer got on so well because each saw in the other the skills and intelligence necessary to fulfill their common goal, the early completion and use of the bomb. The bomb in fact would be the route to immortality for the both of them.

They treated each other in special ways. Oppenheimer could at times be sarcastic, even cruel, with students or colleagues who could not keep up with his quick mind. Not so with Groves. He patiently answered whatever query the general asked. On Groves's part, he treated Oppenheimer delicately, like a fine instrument that needed to be played just right. Groves's normal approach with most of his subordinates was to push them as hard as he could. The pressure was a test to see what they were made of. The more they took, the tougher they were, and the more they would produce. The good ones would make it through; those that did not would be transferred, demoted, or replaced. The general saw that this approach would not work with Oppenheimer. Some men if pushed too hard will break.

Time does not permit in this brief talk to detail his relationships with many of the other scientists. The positive ones with the ever-enthusiastic Ernest Lawrence and with Arthur Compton, who like Groves, was the son of a Presbyterian minister, or with James Chadwick, head of the British mission. A fuller treatment would also have to look at that period immediately after the War in the Fall of 1945 when Groves' reputation went sharply into decline, caused in part by the rise of the scientists' movement and the battle over civilian–military control of the atom that led to the Atomic Energy Act of 1946.

Science in the Service of the State: The Cautionary Tale of Robert Oppenheimer

Gregg Herken
Smithsonian Institution

The notion that scientists have no "proprietary rights" to say what should be done with their inventions was expressed by Robert Oppenheimer a month before the explosion of the first atomic bomb. A month after two atomic bombs had been dropped on Japan, not even Oppenheimer believed in this prescription. In 1949, Oppenheimer would oppose development of the hydrogen bomb on both practical and ethical grounds. The loyalty hearing that took place five years later suggests that science in the service of the state bears a potential cost, for both sides.

Some fifty years ago, when asked what impact the Cold War had had upon his discipline, Princeton physicist Henry DeWolf Smyth reportedly replied: "Secrecy, scientists who will take orders, and big equipment." Two of the things that Smyth mentioned — secrecy and big equipment — are undoubted legacies of the Cold War. But the third — the willingness of scientists to take orders — is a more problematic and complicated tale, precisely because of the Cold War, and because of the particular experience of one scientist who answered the summons: Robert Oppenheimer.

J. Robert Oppenheimer and General Leslie Groves at Ground Zero (9/45).

When he was chosen by General Groves to direct the Los Alamos laboratory in the fall of 1942, Oppenheimer showed no hesitation about working on the atomic bomb. Indeed, after "Oppie" came under suspicion by Army counter-intelligence for his prewar left-wing views, it was Oppenheimer's ambition, not his patriotism, which convinced Groves's head of security, John Lansdale, that Oppie would not and could not be a spy for the Russians.

Near the end of the war, when Oppenheimer was chosen to serve on a scientific panel — with Enrico Fermi, Ernest Lawrence, and Arthur Compton — to advise on the use of the bomb, it was Oppenheimer, the head of the panel, who argued that there was no practical alternative to military use of the weapon against Japan. But as Oppenheimer also wrote to Secretary of War Henry Stimson at that time:

> "With regard to these general aspects of the use of atomic energy, it is clear that we, as scientific men, have no proprietary rights. It is true that we are among the few citizens who have had occasion to give thoughtful consideration to these problems during the past few years. We have, however, no claim to special competence in solving the political, social, and military problems which are presented by the advent of atomic power."

Even at the time that Oppenheimer wrote those words, his was probably already a minority view among scientists working on the project. Leo Szilard obviously felt no such compunction about telling the government what it should do when he passed around his petition at the University of Chicago. Edward Teller claims that he was sympathetic to Szilard's appeal but that Oppenheimer — who reminded Teller that it was not the job of scientists to decide how the bomb was used — forbade him from circulating the petition at Los Alamos, as Szilard had requested.

Within weeks of the bombing of Hiroshima and Nagasaki, Oppenheimer himself had reversed his stand on the role of scientists. By September 1945, Oppie was using his status as the "father" of the atomic bomb to importune top officials in the Truman administration and even Truman himself on behalf of his new-found cause: the international control of atomic energy.

Arguably the most famous, or notorious, instance of scientists speaking the truth to power came some fours years later, in October 1949, when Oppenheimer, then chairman of the Atomic Energy Commission's General Advisory Committee, drafted the GAC's majority report, urging that the nation *not* proceed with a crash effort to develop the hydrogen super-bomb — which the committee described as potentially "a weapon of genocide."

As we know, the GAC's advice was ignored, and a few years later — after the prototype super-bomb had been successfully tested — Oppenheimer was brought to account for his views. Although the impetus behind the revocation of Oppenheimer's security clearance and the resulting loyalty hearing concerned lies that Oppie had told to Army security agents during the war, the evidence now available in AEC and FBI files leaves little doubt that the real motivation behind the Oppenheimer trial — and its verdict — was Oppie's so-called "failure to enthuse" over the hydrogen bomb. The intent, therefore, was not only to end Oppenheimer's influence as a science adviser — which was essentially at an end by this time anyway — but to "unfrock" Oppenheimer before his peers.

With one dissenting vote, the AEC declared Oppenheimer a loyal citizen, but stripped him of his security clearance, just one day before it was due to expire. The verdict had two effects: it made Oppenheimer a martyr in the scientific community, and it was the start of Oppie's academic exile at Princeton, which would endure for nearly a decade. The message that the state seemed to be sending to its scientists in the Oppenheimer case was, "we value your necessary inventions, but not your unwanted advice."

Oppenheimer would be partially rehabilitated in 1963, when he was awarded the Fermi medal, but the Kennedy administration pointedly refused to reinstate his security clearance on that occasion, lest the old controversy be revived. Although Oppie died in 1967, his ghost is still very much with us — as Edward Teller reminded in his memoirs, published just last year. Forty years after the infamous loyalty hearing, Teller would attribute the difficulty he had in recruiting scientists to work on Ronald Reagan's Strategic Defense Initiative to the lingering aftermath of the Oppenheimer case.

Today, nearly sixty years after the Oppenheimer trial, secrecy is with us more than ever; big equipment as well. But the era when scientists were blindly willing to follow orders — even in wartime — may well be over.

Leo Szilard: Baiting Brass Hats

William Lanouette
Writer and public policy analyst

Physicist Leo Szilard was first to conceive the nuclear chain reaction, first to impose secrecy on nuclear research, first to prompt the United States to create A-bombs, and also first to urge wartime and post-war control of these new weapons. Szilard's prescience raised fateful questions about the bomb years before it was tested or used. Yet in Szilard's view the most powerful weapon to result from the Manhattan Project was not the A-bomb but the "SECRET" stamp.

On Monday morning, March 6th 1944, Leo Szilard rode in a taxi from the Wardman Park Hotel to this building, followed by federal agents. Here Szilard called on Vannevar Bush, Carnegie Institution president and director of the wartime Office of Scientific Research and Development. As Bush later noted, their "long conference" was "quite informal and frank" and lasted "practically all day." Szilard had been nagging Bush by letter for almost two years and came here to complain — in person — about how badly America's work on the A-bomb was being managed. Always feisty and persistent, Szilard criticized the "compartmentalization" of sites and scientists imposed by General Leslie Groves, the Manhattan Project's military director. Then Szilard urged that the United States work for post-war international control of a weapon that would not be tested for more than a year.

Groves saw Szilard as disruptive and probably subversive; it was on his orders that Szilard was being followed here. And Szilard saw Groves as officious and incompetent. Both men were wrong. For both men, in very different ways, were essential to the Manhattan Project's success. Without Szilard's tireless efforts at the beginning, and without Groves's relentless race at the end, it is unlikely that nuclear weapons would have been ready in time for use by the end of World War II.

Born in Budapest in 1898, Szilard studied physics in Berlin in the 1920s where he befriended Albert Einstein, and where the two co-designed an electromagnetic refrigerator pump. Szilard fled Nazi Germany in 1933, and that fall, in London, he first conceived the nuclear chain reaction. Always prescient about science and politics, Szilard foresaw both nuclear power and nuclear weapons. He proffered his chain-reaction patent to General Electric for power plants, and to the British Army for bombs. Both had no use for his ideas, but finally the Admiralty classified Szilard's chain-reaction patent as a military secret, lest German scientists realize its potential.

Leo Szilard and Ernest O. Lawrence at APS Meeting in Washington, April 27, 1935 at the Department of Terrestrial Magnetism. (Courtesy of Science Service.)

Although Szilard proudly claimed that his favorite pastime was "baiting brass hats," he strove repeatedly to collaborate with the military in Britain and the United States. And the secrecy he later decried was, in fact, another one of his own inventions. At Columbia University in 1939, soon after uranium fission was discovered, Szilard angered his fellow scientists by insisting that they keep secret their nuclear research. Szilard urged colleague Enrico Fermi to report on their nuclear fission work to the U.S. Navy, but in Washington Fermi was ridiculed and ignored. That summer, Szilard and Fermi co-designed the world's first nuclear reactor. Then Szilard told Einstein about chain reactions, the process that might give $E = mc^2$ a potent new reality. Einstein's response was simple: "I haven't thought of that at all." Szilard proposed and drafted a letter Einstein signed that warned President Franklin Roosevelt about German nuclear advances and led to the U.S. bomb program.

"If the A-bomb project could have been run on ideas alone," the Nobel laureate in physics Eugene Wigner said, "no one but Leo Szilard would have been needed." But it took much more than ideas. President Roosevelt responded to Einstein's letter by creating a Uranium Committee; and as a member of it Szilard urged Army and Navy ordnance officers to fund nuclear research. They were skeptical, and in 1940, when Army intelligence investigated Fermi and Szilard for a grant to support their reactor research at Columbia it was recommended that neither man could be trusted with the "secret work." Ironically, the only secret work then under way in this country was in their two heads.

Fermi and Szilard came from an intellectual milieu in which scientists had the freedom to work as they pleased, at their own pace. Szilard believed the best way to encourage scientific creativity was to give scientists the "leisure" to pursue their wildest ideas. After all, Szilard (like Archimedes) discovered many insights while soaking for hours in his bathtub, his mind pursuing both scientific facts and science fiction. But in industrial America the pace of progress was swifter. America's science establishment took over the Fermi–Szilard research just before the United States entered World War II in December 1941, and by the fall of 1942 the bomb project came under the firm control of General Groves's Manhattan Engineer District: industrial in scale, international in scope, and organized with military precision. Groves even wanted the scientists to wear uniforms, although some were not yet U.S. citizens.

By September 1942, nearly three months before a controlled nuclear chain reaction was first achieved, Szilard had already begun musing about the terrors of a post-war nuclear arms race, and this grim prospect he raised with Bush on the day he came here in March of 1944. Szilard warned Bush that only an international control scheme could avert nuclear catastrophe. And, in what may be the first suggestion of a preemptive strike to prevent the spread of nuclear weapons, Szilard proposed controlling all world uranium deposits "if necessary by force...." Szilard predicted, ominously, that "it will hardly be possible to get political action along that line unless high efficiency atomic bombs have actually been used in this war and the fact of their destructive power has deeply penetrated the mind of the public." Paradoxically, Szilard feared that the bomb he did not want to use to win the war must be used to win the peace.

For many of Szilard's points Bush was sympathetic; he had been brainstorming independently about post-war problems with James B. Conant of the National Defense Research Committee. In May 1944, Conant wrote Bush a memo about "Some Thoughts on International Control of Atomic Energy" that posed two stark alternatives: an atomic arms race and "in the next war destruction of civilization" or "a scheme to remove atomic energy from the field of conflict." In September 1944, Bush and Conant sent a memo to Secretary of War Henry L. Stimson on "Salient Points Concerning Future International Handling of Subject of Atomic Bombs" with their "strong recommendation" to plan for "complete disclosure of the history of the development and all but the manufacturing and military details on the bombs as soon as the first bomb has been demonstrated" — either over Japan or the US. They warned that partial secrecy might prompt a nuclear arms race by Russia and others. And they called for "free interchange of all scientific information."

Leo Szilard (1898–1964).
(Cartoon credit: Robert Grossman.)

Frequently Szilard's views on the post-war control of nuclear weapons were not so different from those held by the civilian leaders in the U.S. bomb program. What was different was Szilard's status in the hierarchy of American science and his ironic and paradoxical views. Szilard believed that he and other emigre scientists had raced to build the A-bomb as a *defensive* weapon against Hitler, not as an *offensive* weapon against Japan — or perhaps even Russia. So, in the spring of 1945 Szilard tried three ways to prevent the atomic bombing of Japanese cities and advocated, instead, a demonstration of this awesome new weapon.

First, in March Szilard drafted another Einstein letter to President Roosevelt, introducing Szilard as someone with important ideas about post-war policies. But Roosevelt died in April before receiving the letter.

Second, in May Szilard called on the Truman White House with the Einstein letter and a detailed memo on post-war control policies. But the president sent Szilard to South Carolina to visit James F. Byrnes, who would soon be named Secretary of State. There Szilard argued against using the bomb and for post-war controls, but to Byrnes this new weapon was just the thing he sought to make the Russians "more manageable."

Third, in June Szilard helped draft the Franck Committee report in which Manhattan Project scientists proposed demonstrating the bomb. This was also ignored in Washington, leaving Szilard a last, symbolic effort — a scientists' petition to President Truman. Szilard began drafting his petition in the first week of July, and sent it up the Army chain-of-command on July 17th, the day after the A-bomb was first tested. To undermine the petition, Groves's staff polled the scientists but discovered that 83% favored a demonstration. In all, 155 Manhattan Project scientists signed versions of the petition, which urged the president to weigh moral and diplomatic issues before using the bomb. But on its way to the White House Szilard's petition was pigeonholed by Groves's staff until after A-bombs destroyed Hiroshima and Nagasaki in August 1945.

Despite its horrors, Szilard said that Hiroshima brought him "a feeling of relief" because "suddenly the secrecy was dropped" and the scientists could speak out. Eager to present his post-war control schemes to the public and policymakers, Szilard asked permission to publish his petition to Truman in *Science* magazine. But General Groves had it classified "SECRET." The scientists' petition was not finally declassified until 1961, and was not published until 1963, the year before Szilard's death. In Szilard's view, the most powerful weapon to result from the Manhattan Project was not the A-bomb but the "SECRET" stamp.

Chapter 5: Speaking from Experience

SEDs at Los Alamos: A Personal Memoir[40]

Benjamin Bederson
Professor of Physics Emeritus at NYU and
Editor-in-Chief Emeritus, American Physical Society

The role played by the common U.S. soldier in the development of atomic weapons during World War II is not generally appreciated. Early in the history of the Manhattan Project, the U.S. Army decided to tap the vast pool of GIs possessing scientific and technical backgrounds who were serving in it, mostly as draftees. These soldiers were assigned to an entity called the Special Engineering Detachment, and hence were known as "SEDs." Their ranks also included skilled mechanics, machinists, and electronics technicians. At its peak in 1945 about 1,800 SEDs were working, mainly at the principal Manhattan Project sites at Los Alamos and Oak Ridge.

The main role of the SEDs was to act as assistants, something like graduate students, to the senior scientists who by then were arriving at Los Alamos and Oak Ridge in large numbers. The SEDs were assigned to many individual research projects that would eventually culminate in the successful design and construction of the two atomic bombs, the "Little Boy" and the "Fat Man," as well as to the various technical shops. Often little distinction was made between people in and out of uniform, although the former were subject to army regulations and discipline, as well as to army salaries rather than civilian ones.

As work progressed, many of these soldiers assumed positions of considerable importance while still being subject to normal army routines, tempered somewhat by the rather loose discipline that prevailed, particularly toward the end of the war.

[40] The entire article may be obtained from Kirkhauser Verlag AG, P.O. Box 133, CH-4010 Basel/Switzerland; subscriptions@birkhauser.ch or http://www.birkhauser.ch (Vol. 3, 2001).

I entered the army through the draft in 1942, possessed of two-and-one-half years of college (City College of New York) as a physics major. At some time in 1943 I had found myself happily back in college at Ohio State University taking an electrical engineering course courtesy of the Army Specialized Training Program (ASTP). This program was intended to teach technical skills to soldiers for an army that was experiencing ever increasing demands on such skills in fighting a modern war. Just as I was completing this course in January 1944, the Army announced that it was going to abandon the ASTP because of increasing demand for combat troops in Europe and the Pacific.

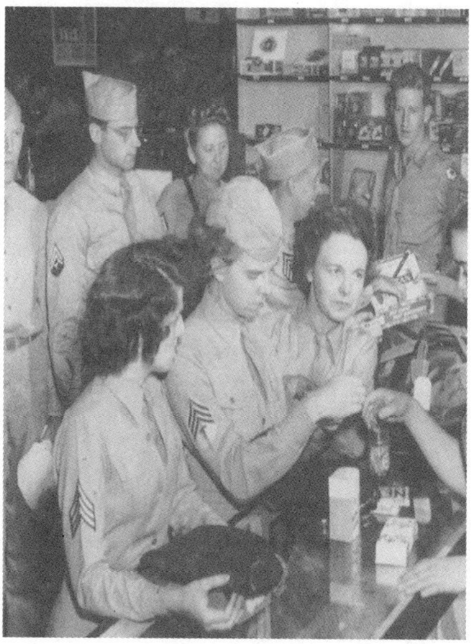

Army Post Exchange at Oak Ridge. (The Department of Energy, "The Manhattan Project," 2001.)

Coincidently, at that moment my commanding officer asked me if I would be interested in being interviewed for a new project, called the Manhattan Project, where my physics and engineering training, such as they were at the time, might come in handy. And, he remarked, this might get me back to my beloved Manhattan, of whose affection I had made no secret in Columbus, Ohio. Needless to say, I jumped at the opportunity, and shortly thereafter was interviewed by a visiting board of three civilians. They asked rather peculiar questions, I thought at the time, consisting mainly of elementary physics questions, for example, about Newton's laws, and about my career interests.

A few days later I received orders, marked Secret, along with a train ticket, to proceed to a town called Knoxville, Tennessee, to be met there by a car that would take me to another town called Oak Ridge. On the train I met several other GIs who also had received the same orders. We arrived in Oak Ridge to discover a city in the throes of heavy construction.

There was orange-red mud everywhere, and a number of tall buildings with a peculiar look.

If I hadn't known better I would have sworn they were moonshine factories. In view of their location in the hills of Tennessee, I first thought that the government was secretly manufacturing Tennessee sour mash on a huge scale — perhaps to drop huge vats of it over Europe to disable the German troops. It turned out that the plants were indeed distillers, only not for whiskey but rather for the gaseous diffusion separation of uranium-235 from uranium-238. Of course, I didn't discover this until later. I was assigned to the Special Engineering Detachment, the SED, and I remained in it until my discharge in January 1946.

Something unusual was obviously going on at Oak Ridge. While we were quartered in barracks, they were cleaned and our beds were made by local young girls! No KP, latrine duty, or even drill — not a typical army experience. Instead, we were given a series of tests and interviews. After about a week I received new shipping orders, this time to report to an address in Santa Fe, New Mexico.

Again I traveled by civilian train, to a town called Lamy. That was when I discovered that the Atchison, Topeka and Santa Fe railroad did not (and still doesn't) go to Santa Fe. At Lamy I was met by an army sedan and taken to an address in Santa Fe, the now famous 109 E. Palace Avenue. It was a store front on the main Plaza, where just about everyone going to Los Alamos reported (to Dorothy McKibbin) before heading up "the hill."

Before long I found myself in another army sedan, which drove me north, then west, up a tortuous road along the side of a mesa, without guard rails at the time. I was assigned to a barracks, and then to a project. As for army life, we did indeed partake of this during the time when we weren't actually working. We had to undergo hated calisthenics in the early morning, and traditional Saturday morning inspection, for example — but no KP! Comparing notes we discovered that most of us had something in common — our educational or training backgrounds in science, particularly in physics and chemistry. Also scattered among us were machinists, including my neighbor in the next bunk, David Greenglass.

My first supervisor turned out to be a British physicist by the name of Philip B. Moon. He arrived at Los Alamos just weeks after I did, as one member of the group of British scientists who had by agreement joined forces with American scientists to work on the Manhattan Project, rather than working independently in Great Britain. He was the first practicing physicist I had ever met (apart from my City College professors), and what an introduction it was! Phil, as he insisted on being called, was an amazingly sloppy dresser. The first thing I noticed about him was that he wore a tie for a belt. And he perpetually had a cigarette dangling from his lips, with one eye closed to keep the smoke out, and with ashes forever falling randomly about him.

Much later I learned that he was one of the scientists who had performed very early work on the neutron, after its discovery by Chadwick. We became good friends — I was greatly taken by his wry sense of humor, and we corresponded with each other for a while after the war. He had a wife, Winifred, who was the quintessential English matron, very tall and broad, with an accent I could never understand, but warm and friendly. Much later, when I had helped organize the so-called "Mushroom Society," a club consisting of myself, Norman Greenspan, and occasionally Ted Hall, Winifred and Phil would on occasion come as invited guests to listen to our classical record collection in a small office where we had installed a homemade hi-fidelity system.

My first assignment at Los Alamos under Professor Moon involved the testing of "strain gauges." These are thin wires attached to a sample whose ability to withstand stress is determined by monitoring the change in resistance of the wires as the sample is stretched. My job consisted mainly of attaching these gauges to cylindrical hollow tubes containing explosives, which were attached at first to "primacord." Primacord consists of ropes of explosives, very handy for use by sappers to blow up bridges and railroad lines. A bit later I was given electrically activated detonators to use in the tests. These were very tricky since they tended to go off without warning owing to static electricity, which was easy to create if the electrical connections were not properly grounded. And in fact such a detonator did explode in the hand of my army buddy working next to me on an experiment — John McNamara, a cheerful bulky fellow from Wagoner, Oklahoma. He was severely injured, although he fortunately recovered.

In a month or two I was called to a small meeting of SEDs who, like myself, were working on various aspects of explosives. At the meeting we were greeted by the head of the Explosives Division, George B. Kistiakowsky. "Kisty" was a professor of chemistry at Harvard, one of the most distinguished chemists in the world, as I was to find out later. He had a strong Russian accent, and was very approachable and good-natured. The purpose of the meeting was to let the GIs know what was going on at Los Alamos. He laid it all out, from beginning to end.

George B. Kistiakowsky. (Richard Rhodes, *The Making of the Atomic Bomb* (New York: Simon & Schuster, 1986).)

The story that people only knew exactly what they needed to know to do their job was simply untrue. Kistiakowsky explained nuclear fission, critical mass, and the implosion concept among other things. None of this was needed for my job, which, it turned out, was to help create something called Jumbo. Jumbo was a huge cylindrical container into which was to be placed the first "Fat Man" to be tested. Fat Man was the name of the implosion bomb (generally referred to as the "gadget"), and Jumbo was intended to contain the radioactive material if the nuclear explosion failed but the TNT explosive lenses did enough damage to spread deadly radioactivity from the unfissioned plutonium. Jumbo was to prevent this radioactivity from spreading all over the landscape — and, incidentally, making it possible to recover the unspent plutonium for another try. That is why I was testing containers!

Everything fell into place with Kistiakowsky's revelations, from the mysterious distillation plants in Oak Ridge to the overwhelming secrecy of the entire project. The only thing I had cause to be miffed about was my faded hope that the Manhattan Project would get me back to New York. Still, the thought that somehow I had landed in the middle of what was certainly a historic enterprise was exhilarating and inspiring.

At this time I had what was called a "blue badge" — a sort of second-class admission ticket. It allowed me to wander around all of Los Alamos except the "Tech Area" — the collection of buildings guarded by MPs (Military Police) where the main work on the bomb was clearly taking place. I could drive out to "Two Mile Mesa," where the strain-gauge tests were being performed, clearly an assignment that did not possess the highest priority. [Up to this time I had never driven a car. This was easily remedied, however. I simply signed out a car from the motor pool and taught myself how to drive.]

I did not know it at the time but during this period I was being given a security check. Army security people (not the FBI) went to my neighborhood in the Bronx, Allerton Avenue, and interviewed people who knew me there. These included the superintendent of my apartment house as well as my two closest friends at the time, both of whom gave me sterling recommendations and assured the investigators that my politics were above reproach. I am eternally grateful to them for telling those small white lies, particularly the one in response to the question about the "Coops"— the communist cooperative houses where I did indeed hang out, and where my early political beliefs were formed. Apparently the investigators also did not view my stay in the U.S.S.R. in 1932 at the age of 12 with my family for about eight months as being particularly sinister.

In any case, it wasn't very long before I received word that my security investigation had been completed. I was given a white badge, the prestigious symbol of complete acceptance into Los Alamos society, which allowed me almost unlimited access to the Tech area. I immediately was permitted to attend the general Tuesday evening talks, during which various leaders gave progress reports on their work. Thus I was able to gain a pretty complete picture of most of the important aspects of the bomb work.

The most memorable of these talks was also the first one I attended, not more than two or three days after receiving my white badge. The talk was given by Enrico Fermi. Fermi was an affable, obviously brilliant physicist with a strong Italian accent. He sounded just like some of the store owners on Allerton Avenue. But what he spoke about was a far cry from salamis and mozzarella cheese. He didn't even speak about the fission project. In his mind he had already assumed that the project would be successful, and he was thinking ahead, to the hydrogen bomb.

He gave a fully developed talk on the expected yield of the fusion reaction designed as a bomb, and he offered some thoughts on how it might be designed. This was probably in March or early April 1944.

Later talks were given by various Los Alamos luminaries, including Edward Teller, Philip Morrison, "Nicholas Baker" — that is, Niels Bohr — and many others. Simply attending these talks as a low-ranking GI was an incredible piece of luck and a wonderful privilege that helped me define my later career.

In all of the time I was at Los Alamos I met its two principal leaders exactly once each. Shortly before Christmas 1944, a handful of SEDs were invited to meet General Leslie R. Groves. While I was expecting something a bit more world shaking at this momentous event (for a corporal), General Groves revealed his purpose in arranging the meeting. It was, he said, to urge us to write home to our parents at Christmas time. You have no idea, he said, how much this would mean to them. At the meeting he also asked whether we had any complaints or suggestions on how life could be made better for us. Only one SED spoke up — he was very unhappy that the Army did not supply us with adequate baseball equipment. We needed more balls, gloves, and bats. General Groves promised to look into it, and we adjourned.

Author Benjamin Bederson (center) and others on Tinian. (Bederson, "SEDs at Los Alamos," 2001.)

The day before I was to take off for Tinian [an island in the South Pacific where the Air Force had a large base], the group that was to travel on the Green Hornet assembled in Dr. Oppenheimer's office. He gave us a rousing speech, telling us how important our mission was, he shook our hands, wished us the best of luck, and we adjourned.

Some Experiences at the Met. Lab and What Could Be Learned from a Highly Successful and Challenging Project

Jerome Karle
Naval Research Laboratory

Jerome Karle

This article concerns some experiences of an employee of the Manhattan Project on the level of a laboratory experimenter. It also concerns some suggestions regarding how science can progress in an optimal fashion as implied by experiences in the Manhattan Project.

In the summer of 1943, I finished my Ph.D. thesis on the determination of the structures of gaseous molecules by use of electron diffraction. My Ph.D. adviser was Professor Lawrence O. Brockway, one of the first graduate students of Linus Pauling. Pauling had obtained the idea of electron diffraction from Herman Mark's laboratory on a visit to Germany.

Lawrence Brockway (*left*), thesis advisor, with Isabella and Jerome Karle.

I had some experience with research on behalf of the war effort while I was a graduate student. I assembled an electron diffraction apparatus designed for investigating surfaces. It was used in a project that Brockway had arranged with the Ford Motor Company to investigate the nature of aluminum surfaces. The chemical irregularities of the surfaces caused difficulties in welding applications. At that time, the Ford Motor Company was manufacturing bombing planes.

After I completed my thesis, I quickly found employment with the Manhattan Project at the University of Chicago. In Chicago the project was called the Metallurgy or Met. Lab. A few months later, my wife finished her thesis and joined me in Chicago. We lived on 52nd street a few blocks from the IC (Illinois Central Railroad) and Lake Michigan. Bicycles brought us back and forth to work in a building which was called New Chem., a short walk from the main campus of the University of Chicago.

When I first arrived for work, I was introduced to a very busy group of young people. The personnel were largely people in their twenties and early thirties. Glenn Seaborg, in his early thirties, was the head of the project in New Chem. There was one level of managers between me and Glenn Seaborg. I had very little contact with the person in charge of the group that I joined.

I immediately found out that my project was to try to find a way to remove oxygen from the oxides of uranium and plutonium. Before I came to Chicago, another person had built an apparatus for carrying out a possibly suitable experiment. Its main feature was a way to create atomic hydrogen in a vessel which contained oxidized uranium or plutonium. It was hoped that the atomic hydrogen would remove the oxygen. A variety of attempts with uranium over several months was not successful. I then began to think about alternative approaches.

I was aware of another method that could be tried. The idea was to use a centrifuge that supported a special container in one of its cups. The container needed to be able to be heated to high temperatures, about 1,100 degrees Fahrenheit, as I recall. It was necessary to modify an ordinary centrifuge. There were heater wires and wires to deliver information concerning the temperatures obtained. These needed to be operated through slip rings, evidently because the centrifuge cups rotate and the recorders do not. I was able to make a suitable instrument without much trouble.

I tested the physical features of the instrument several times with the use of uranium in the absence of a reducing agent. At this point the main interest was in plutonium. When I was ready, a tiny amount of oxidized plutonium was made available to me. This was placed in a container made of stainless steel along with a substance that has a high affinity for oxygen. The container was then sealed with a cap also made of stainless steel. People in a number of neighboring rooms were asked to leave in case of an accident and I was ready to carry out the experiment with the modified centrifuge placed in a hood.

Everything went very smoothly, as expected from the preliminary experiments that used uranium. When the container was opened, there was a bright shiny portion that was removed for further examination with a microscope. It was generally thought that the experiment was at least partially successful and further experimentation with this technique was turned over to others to pursue.

Not too much later, my thesis professor contacted me, asking me to return to the University of Michigan and help him with a Navy project. Our activities were beginning to slow down at New Chem. and so we decided to return to Michigan.

The atmosphere at New Chem. concerning the objective of making an atomic bomb was both serious and concerned. The concern was subdued however because of the continuing possibility that Germany may succeed in making atomic bombs.

We had the opportunity to meet distinguished people and make some lifelong friendships. One gentleman with whom I had several pleasant discussions was James Franck, Nobel Laureate in Physics in 1925. William Zachariasen of the Physics Department at Chicago was a distinguished crystallographer with whom we had many contacts during our lives.

I saw Enrico Fermi many times sitting on the lawn of the campus with some of his students joining them eating a bag lunch. I never had a conversation with him. I did hear a very interesting talk that he gave explaining his need to build the pile beneath the stands in the sports stadium. He was not sure whether the energetic process would be self-sustaining. He said that his calculations were not accurate enough to guarantee it. So, verifying the process was one of the main motivations for making the pile.

Another person with whom we were in contact over the years was Glenn Seaborg, the leader of the program at New Chem. and Nobel Laureate in 1951. Being a member of the Royal Swedish Academy of Sciences, he often went to the yearly Nobel Prize events. He made sure to be present when I received the Nobel Prize in 1985.

It is quite apparent that the Manhattan Project was a very difficult undertaking characterized by many new areas of research and the requirements of profound insight and understanding of the physics and chemistry involved. Nevertheless, the project reached its objectives in a relatively brief period of time. With regard to scientific research institutions the question arises, what should we have learned from the Manhattan Project? We should have learned the following:

1. Excellent leadership should always be sought. Leaders should understand well the scientific developments that are worth pursuing. Their scientific knowledge and background should be such that promising paths for research can be recognized and encouraged. Management skills without a deep understanding of scientific research and how it proceeds is an invitation to disaster.
2. Employ the best people who can be found.
3. Minimize interruptions and intrusions on a scientist's research program. By imposing endless administrative notices, report requirements and by forcing scientists to write numerous proposals to support their research, much precious time is wasted and valuable thoughts can be lost. Less than total direct funding of federal and university institutions has helped to generate the present state of science in the United States, namely, less productivity from publicly funded science than is possible and the disinterest of many talented American students in a career in science. The sources of this waste should be eliminated.

It should be noted that public money invested in scientific research and technology has brought considerable economic return to the public. The economic return on federally supported research was the subject of a report by the Council on Economic Advisors to President Clinton. It was released in November 1995 and documents high returns to the economy. It could be more for the same investment. In addition, potential weakening of the scientific profession could be avoided.

My First Professional Assignment

Isabella Karle
Naval Research Laboratory

Experiences of a new doctoral recipient (physical chemistry) in synthesizing $PuCl_3$, the environment at the Met Laboratory and the future and present accomplishments of the personnel are described.

Isabella Karle

In December of 1943, I celebrated my 22nd birthday. I also completed all requirements for a Ph.D. degree in physical chemistry at the University of Michigan, packed my few belongings and took the train to Chicago where I joined my husband (Jerome Karle). He had already been employed for several months on a mysterious project at the University of Chicago. I also received an offer of employment for the same project which I accepted — sight unseen by both employer and employee. There was no application for the job. There was no personal contact with me and no information about the nature of the job. When I asked my husband about what I might be doing, all he said was that I will be amazed.

Immediately after New Year's Day in 1944, I presented myself at the University of Chicago where I filled out a few forms, was fingerprinted, and was told that I had to use my married name (I had retained my maiden name through graduate school). Then I was sent to a barracks-like building which was named "New Chemistry" and met my immediate supervisor, Norman Davidson. I already knew Norman since I shared an office with him when he was a postdoctoral appointee for a brief time with Prof. Lawrence Brockway, my thesis advisor. He left Michigan suddenly and I did not know why or where he had gone until I met him again at the Met. Lab.

There was very little indoctrination or introduction to my new job. I was told that plutonium (Pu) was a new element — not to be found on the periodic table and that Norman Davidson was in charge of a group of 6–7 people whose task it was to synthesize plutonium halides by vapor phase procedures. My specific task was to synthesize pure plutonium chloride. The starting material was crude plutonium dioxide in the form of greasy yellow flakes as it was delivered from Oak Ridge, Tennessee. The ultimate objective was to produce pure Pu metal, by another group. However, it was deemed necessary to learn about plutonium chemistry and it was thought that one of the plutonium halides would be a good intermediate in the process of producing the metal.

The only practical information imparted to me was that plutonium may be similar to uranium in many of its reactions. Accordingly, I spent the first week in the library to learn all I could about the chemistry of uranium. Then I started to assemble a glass vacuum line in which the reactions were to take place. During my graduate studies I had acquired considerable experience with glass vacuum line technology and the associated glass blowing techniques. Since I expected that the reactions should occur at ~800°C, I needed a removable insert in the vacuum line made of quartz that would be the reaction chamber. My glass blowing techniques also worked for handling quartz with a hydrogen torch and extremely dark glasses. Furthermore to reach the high temperatures required for the reaction, I used a block of copper metal that was hollowed out sufficiently to surround the quartz reaction chamber. The high temperature was attained by heating the copper block with a torch until the metal almost glowed. The experiment consisted of passing a chlorine containing organic vapor over the plutonium dioxide for about an hour. In separate trials I tried a number of chlorinated compounds, all of which produced plutonium chloride, however, carbon tetrachloride appeared to be one of the most suitable. After the cooling of the copper furnace and opening it, brilliant green crystals were found in the place of the grungy yellow oxide.

To prove that the crystals were plutonium chloride, I transferred them into a vial, using a homemade dry box for the transfer. The dry box was designed and constructed in the laboratory by one of my colleagues. I then put the vial in my pocket and walked across the campus to the physics department where Prof. W. Zachariasen performed X-ray crystal structure analyses on the crystalline sample, which showed that the new compound was plutonium trichloride, not the tetrachloride that occurred for uranium.

I repeated the experiment a number of times with different conditions and different chlorinating compounds. Then the security department found out about my transferring samples to the physics department by myself. After that, I had two uniformed guards marching with me, probably alerting individuals in view of the campus that there was probably something unusual going on.

Another interesting event concerning security occurred when the secretary for our department could not come to work one morning. Everyone's notebooks were locked in her safe every night when we left the laboratory. Since she had the only key, the scientists opened the safe by removing the pins in the door hinges by using only a screw driver and a wrench.

Glenn Seaborg was concerned about the effects of radiation with respect to the alpha particles emitted by the plutonium. Although the total amount of plutonium in the entire laboratory was relatively small, perhaps a maximum of 100 mg at any one time, a number of precautions were instituted. We had our hands and feet counted for radiation and urine examined once a week. Every day we were given a chocolate-flavored calcium tablet the size of a 50 cent coin. Presumably, an excess of calcium ions would prevent plutonium ions from depositing in the bones. It was also recommended that the pumps for evacuating the room-sized hoods be increased in pumping power.

There were however unexpected events. I will describe one of them. This concerned the Coca Cola machine. It was a style of machine that dropped a paper cup, which was then filled with carbonated water and coca cola syrup. The man who came to service the machine at our lunch time forgot to bring his hose for filling the syrup reservoir. He walked into the neighboring laboratory where wet chemistry was being performed and borrowed a rubber hose from an aspirator, filled the reservoir with the syrup, returned the hose and left. Some time after lunch a technician was carrying an alpha counter and noticed that the meter went off the scale as he passed the Coke machine. By the next day, the Coke machine was replaced with one that dispensed bottles rather than liquids. We never did know how many, if any, employees drank the radioactive Coca Cola.

Recently it was with great pleasure that I found the names of almost all of my contemporaries at Chicago whose names I could remember, in the latest issue of American Men and Women in Science. This meant that not only did we survive our experiences with plutonium, but also that we had long and productive careers in science. We are all at least 80 years old, and some are more than 90 years old. Perhaps, exposure to a little radioactivity had a positive effect.

I should like to mention the names of my colleagues and briefly their future scientific accomplishments. In the plutonium halide group: Norman Davidson: Distinguished Professor of chemical biology, Cal Tech, National Medal of Science; Joseph Katz, Distinguished Senior Scientist, Argonne National Laboratory, Member National Academy of Science; Roy Heath, Dean of School of Graduate Studies, Northern Michigan University; French Hagemann, Associate Division Director, Argonne National Laboratory; Michael Wolf, Research Leader in cereals, USDA, Peoria, Illinois; Earl Hyde, Senior Staff Scientist in nuclear chemistry, Lawrence Berkeley Laboratory. Other colleagues in nearly laboratories: Lawrence Bartell, Distinguished Professor in physical chemistry, University of Michigan; Jerome Karle, Chief Scientist, Laboratory for the Structure of Matter, Naval Research Laboratory, Washington, D.C., Nobel Prize in Chemistry; Sidney Katz, Research position at the Illinois Institute of Technology (in Chicago); Alvin Dirksen, Professor in Business, San Francisco State University; Lyle Jensen, Professor (protein crystallography), University of Washington; Winston Manning, Co-discover of Einsteinium and Fermium in 1952, manager at Argonne National Laboratory; Harlan Baumbach, Film processing at Paramount and other movie studios in Hollywood. Two Oscars for Scientific Technical Achievement in Motion Pictures.

The person or persons who selected the scientific staff for the Met. Lab did a remarkable job in a very short time. The selectees came directly from graduate schools, junior faculty members, industries such as Wyandotte Chemical, Oldbury Electro-Chemical Company, and Paramount Motion Picture Studio. The research they performed at the Metallurgical Laboratory was successful and their future careers have been exemplary.

TRIUMPH AND TRAGEDY: THE ODYSSEY OF J. R. OPPENHEIMER — A PERSONAL PERSPECTIVE

Maurice M. Shapiro
Visiting Professor, University of Maryland

Maurice Shapiro (*ca.* 1990)

A conference on the Manhattan Engineer District is sure to include many stories about Los Alamos — not all of them correct. A good deal is likely to be said about J. Robert Oppenheimer, to whom not only the USA but the world owes a great debt for his unique role in shortening and terminating World War II. As one who was privileged to work on "The Hill" in those years, I would like to share with you some of my recollections.

Arthur Holly Compton, who was my mentor at the University of Chicago, recommended Oppenheimer to General Leslie Groves for the leadership of the bomb laboratory. Compton, who was himself involved in pre-war cosmic ray research, admired Oppy for his brilliant work on cosmic ray theory. To many it seemed an unlikely choice — the quintessential ivory tower and otherworldly academic to lead an intensely practical and high pressure project. It turned out, moreover, to be an extremely difficult task that faced daunting problems of physics and engineering. There was also the challenge of assembling an extraordinary community of scientists and engineers, and then persuading the "prima donnas" among them to work harmoniously.

A crucial reason for the Project's timely success was Oppy's wise insistence, over objections of the military, upon non-compartmentalization within the Los Alamos Laboratory. This eight-syllable word was anathema to Gen. Groves, but a *sine qua non* for Oppy, who would probably have resigned if the General had insisted. Let me cite several examples of decisive discoveries and brilliant insights that might not have occurred if a "need-to-know" policy had been in force. In each instance, it was not the result of a specific work assignment that led to the important ideas, or to solutions of difficult problems. Rather it was the fact that every scientist on the project was well informed of the purpose, the basic strategies and their difficulties, and the proposed methodologies and solutions. There was frequent brainstorming, and everyone was encouraged to contribute ideas that might surmount impediments.

My first example is a discovery by Emilio Segre, an early collaborator of Enrico Fermi, who was destined in post-war years to become a Nobel laureate for his detection of the anti-proton. To my knowledge, no one had asked Segre to look into the possibility that some isotope of Plutonium, Pu-239, might undergo spontaneous fission — i.e., fission not induced by neutrons. But that is what Segre found — that Pu-240, which is inevitably produced along with Pu-239 — has such a high rate of spontaneous fission that it would lead to pre-detonation of a critical mass assembly unless the assembly were achieved at a far faster rate than that of the gun assembly using U-235. Since the rate of U-235 production by separation of the uranium isotopes in Oak Ridge was painfully slow, while the production of plutonium at Hanford looked far more promising, the Laboratory embarked on a crash program of learning how to make a viable plutonium bomb.

This brings me to my second example. The assembly of a critical mass of U-235 in a gun barrel occurs on a time scale of milliseconds, while the time in which a Pu bomb must be brought to criticality is of the order of microseconds. How can this be affected? Seth Neddermeyer, who had worked with Carl Anderson in the discovery of the positron, came up with an answer: implosion using high-explosive lenses could squeeze a central shell of Pu into a critical mass sufficiently fast so as to overcome the hazard of pre-detonation through spontaneous fission of the Pu-240. It became evident, however, from many tests using explosive lenses, that there was a severe danger of the production of jets due to Rayleigh-Taylor instabilities; these jets could radically reduce the yield of the bomb. An effective and simple solution was offered by Robert Christy, a one-time student of Oppenheimer's: Why not replace the shell of Pu with a solid, sub-critical sphere that would be less likely to produce the troublesome jets? The implosion would then compress the ball of Pu to criticality. This design, the "Christy gadget" ("gadget" being a euphemism for "bomb"), was a solution to the "jet problem." It is noteworthy that Christy had not been "assigned" to solve this problem, and under a regime of compartmentalization he might not have known of it!

Let me finally relate a few telling personal experiences with J. Robert Oppenheimer. Upon arrival on "The Hill," I was appointed to be a group leader for "delivery problems" in the Division headed by "Deke" Parsons [associate Director of the Laboratory, and later to become the first "atomic admiral"]. I was also to serve him as consultant on hydrodynamics. A few days later, Oppy set up a briefing to acquaint me with my mission in Los Alamos. If I was surprised by this gracious gesture, I was astonished at the roster of scientists who joined us in the Director's office. They included John von Neumann, Parsons, George Kistiakowsky (head of the Explosives Division), Christy (later provost of Cal Tech), William Penney (postwar head of the British atomic energy project), Ed McMillan, Norman Ramsey, and Hans Bethe (the latter three all to become post-war Nobel laureates). In this august company I, the youngest of the lot, felt like a worm. As I wondered why I "deserved" to be initiated into the Project by this distinguished group, I speculated that Oppy may have been treating me as a fellow cosmic-ray worker. In retrospect, I believe the reception I got was an example of Oppy's style of management that earned him the respect and loyalty of his colleagues.

Another encounter illustrates Oppy's constant concern for security — in glaring contradiction to the accusations against him in the post-war security hearings. A group of leading Los Alamos scientists was having lunch in a huge mess hall at the Air Force base in Wendover Field, Utah, site of the B-29 squadron training for the subsequent drop over Japan. From here planes with dummy bombs were dispatched for practice flights and trial drops in which the Norden bombsight was used. As it was a late lunch, we were a solitary group, eating in the center of the large dining hall, which was empty except for us. I mentioned to Oppy, who was sitting across from me, that my mission in Wendover was to meet with Norden bombsight experts. He glared at me. "I really wouldn't use that word!" I felt chastised — notwithstanding that no outsider could possibly have heard us!

On the other hand, a compliment from the Director was to be cherished. This happened to me shortly after the end of the War. Oppy was escorting the members of the Lilienthal Committee around the sites of the Manhattan Engineer District in order to acquaint them with the problems involved in framing a scheme for international control of atomic energy. He agreed to persuade Lilienthal, during his visit at Los Alamos, to address the Association of Los Alamos Scientists (ALAS). I was Chairman of ALAS, so it fell to my lot to introduce the celebrated head of the Tennessee Valley Authority. As I wondered how much to say, it occurred to me that in introducing a great man, the less said, the better. As we were walking out after the meeting, Oppy turned to me and said, "That was just right." I floated on air.

Chapter 6: Lessons of the Manhattan Project for the 21st Century

Then and Now

Maxine Singer
President, Carnegie Institution of Washington

Maxine Singer

Perhaps before I begin my assignment for today I should to tell you how honored and pleased the Carnegie Institution is that this event is being held here, the place where it all got started. Our panelists' charge is to talk about the lessons for today from the Manhattan Project. As the first speaker in the panel, I will try to set the stage by talking about some of the differences between then and now: differences that are relevant to any lessons we might learn.

I can only note a few of those differences. The one overriding difference is that we as a nation, and the international community, are very different than in 1940, partly as a result of what happened in those years. Another thing that's different is the nature of those we assume to be the aggressors. Then, it was primarily nation-states whose goals were clear to them and more or less clear to those they were fighting against. Now, rather than nation-states we have terrorists with complex loyalties and goals, including the short term goal of terror itself. For the United States, then it was primarily our military that was the target of action by an enemy, whereas now, it's largely civilians that we worry about. Therefore rather than primarily a federal government responsibility, we need to think about roles for state and local governments because of the nature of some of the new threats. With respect to technology, the kind of technology that the Axis powers might use was largely predictable, except for the question of nuclear weapons. Now it's clearly quite unpredictable. There are a broad range of weapons or things that can become weapons, some of them obvious, and some of them not-so-obvious. With respect to the world view of our country, at the time of the beginning of the Second World War, the position of the United States was quite clear, and its goals were clear. But now the position of the United States is not quite clear to many, and the goals of our government and our country are also unclear in many respects. With respect to information flow, then we thought that it could be secure. Now, we know that it's unlikely to be secure because of the changes in technology.

What about the availability of scientific and technical research and advice to the government? Then, at the start of the war, there wasn't much research within the government. About 0.6% of the War Department budget in the very late 1930s was for research, and the research that was going on was not particularly innovative. Daniel Kevles, in his book *Physicists: The History of a Scientific Community in Modern America*, has what I think is one of the great put-downs, when he says, with respect to the War Department research in the late 1930s, "it had at least brought devices of World War I to a state of greater sophistication and effectiveness."[41] At the time, the War Department was not particularly open to innovation and new ideas. That was the challenge that was addressed by Vannevar Bush and the Office of Scientific Research and Development (O.S.R.D.). Now, there is a huge government establishment in scientific research, both in-house and outside, through sponsored research. However, it's not clear how well government agencies use that information, share that information, interact with one another, or call upon external expertise. For example, the experience of scientists who have volunteered ideas or efforts to Governor Ridge and the Office of Homeland Security has largely been frustration with the lack of response. The recent brouhaha over the JASONs is another indication that once again, our government is not necessarily as open to ideas from outside the government as it might be.

One of the things that was obviously of concern then was the availability of trained people. The scientific establishment was not nearly the size it is now, and Vannevar Bush and his colleagues had to look around the country to find people. Now, we have an enormous scientific and technical community, but in fact many of the innovative people are not American citizens and questions have been raised about their participation.

---

[41] Daniel J. Kevles, *Physicists: The History of a Scientific Community in Modern America* (Cambridge, MA: Harvard University Press, 1995).

Vannevar Bush, President of the Carnegie Institution of Washington (1939–1955).

At the start of the Second World War there was no formal science advising to the federal government, including the executive branch. That, of course, was the niche to which Vannevar Bush appointed himself. Since 1950, thanks primarily to William T. Golden's efforts and his report "Mobilizing Science for War," we have had (with some ups and downs) a science advisor to the President. Bush and Golden represent an interesting historical phenomenon. Then, a single individual could find a route for important ideas and get significant official people to listen. There is a real question as to whether such routes exist today. Everything from the Einstein letter in 1939, through Bush's memo to F.D.R., and Golden's report to President Truman, addressed matters of great importance and were initiated by single private individuals. It's not clear that we can successfully do that now. Now, our president is surrounded by a vast bureaucracy and by advisors whose primary focus appears to be political.

As we ask ourselves what kind of lessons we can learn from the organization and the origins of the Manhattan Project as a model for effective work at present, we need to keep in mind that we live in a very, very different world. Also, we have to realize that we were lucky that it was Bush and his particular colleagues who assumed leadership in the Manhattan Project, and were accepted without reference to their politics. That model has not always worked very well. For example, the model of a single person as an advisor to a government is also the model that applies to Lysenko, who was successful because he gained the confidence of a powerful individual. The result was disaster for Soviet genetics and biology. So while we think about the advantages of individuals and their convictions and visions, we also have to think about the problems that might arise. I'm going to leave it to those who follow me on this panel to come up with additional models. I hope you will test them against the "then" and "now" that I have described.

The Manhattan Project: Qualitative or Quantitative Change?

Stephen Younger
Director of the Defense Threat Reduction Agency, Department of Defense

I have spent twenty years of my career at the National Laboratories and have recently come to work for the Department of Defense, and for the first twenty years I basically said as I pleased. When I got to the Department of Defense, I found that things needed to be cleared through Policy and became much more complicated. As a compromise, I will tell you that what I am about to say does not represent the views of the United States Government, the Department of Defense, the Secretary of Defense, or the horse he rode in on, but they do represent my views. So having done that, I'd like to share with you two contradictory theses, which I think are both true; because they deal with human beings I think you can hold both at the same time. The first is that the Manhattan Project and the atomic bomb did not represent a qualitative change in human history but represented a quantitative change; and the second thesis is that it represented a qualitative change — and I'll try to talk about both of those.

This symposium comes at an interesting time. A year ago we referred to the period in which we lived as the post-Cold War period. That meant we weren't quite sure what period it was but we were pretty sure it wasn't the Cold War period. Now, after the events of September 11th it's pretty clear what period we're in. It was a shock to the system: It is the fight against terrorism, which I think will take some time to route out because it is that most difficult fight — a sociological rather than military or even political one. Maxine mentioned that the impact of scientists in government to a certain degree has diminished over the years. That is true, but so too has science moved from scholarship to being a special interest in and of itself.

There are science lobbyists just as there are lobbyists for other kinds of projects that become big business. But the Manhattan Project was arguably the first time that science was marshaled on a massive scale and involved expertise which was the cream not only of the planet at that time, but one could argue of the century, or almost of the millennium with the exceptions of Newton, Galileo, Copernicus and a few others.

It was an unbelievable concentration of scientific and engineering expertise that created a weapon so powerful that after its first uses it was not used again. The atomic bomb was, and is, unique in its destructive force, however, it is important to note — and again, there's a thesis/antithesis — it is important to note that it wasn't the first weapon of mass destruction, and I'll give you an early example. As a matter of fact, I'll give you an example from the first millennium B.C., just to show that you can go back quite a ways. When the Assyrians — who lived in northern Iraq, and who were a warlike people and very efficient at it — had a problem with the city-state on their periphery they would attempt to solve it by diplomatic means. If that didn't work they would lay siege to the city. They always won. They would either kill or deport the entire population of the city-state. They would tear down all of the walls and buildings. They would bury the bricks and salt the fields. That entity would cease to exist as a civilization. Sometimes it would take them a year or two; nevertheless, they had the ability to destroy a civilization as such. Nor was the atomic bomb the first challenge in the continuing contrast of the good and evil uses of technology. It's interesting to note that in the book of Genesis, the tree is identified as the tree of the knowledge of good *and* evil. Not good *or* evil but good *and* evil which is interesting because I think all things can be used for good and for ill. A knife can be used to slice bread; it can be used to kill. Medicine can be used to cure disease; it can be used to create disease. So from that standpoint, one might see all of history as a progression of increasingly more complex choices. The challenge of history is not to avoid those choices, but to face them squarely and to make the right and enduring decisions. Nuclear weapons will not go away. One could argue whether they should have been created in the first place, that is an argument of some interest but irrelevant in that they do exist. We don't have the option of putting that demon back into Pandora's box.

With the dawn of the atomic age we had to adjust to a new reality and make decisions about how we would use that new reality. But as strange as it may seem, for a weapon of such force I believe that we chose wisely. Why do I say that? Statistics. If you plot, in a cold-hearted way, the number of people killed in organized warfare over the past two centuries, you will note that it rose at an almost-exponential rate due to the inextricable advances of weapons technology. With the advent of the atomic bomb that stopped. Does that mean that all conflict in the world stopped? No, and I'll address that in a moment. Does it mean that very large-scale destruction through strategic warfare stopped? Yes, for a while.

The atomic bomb was a sufficient shock to the social system that it caused a pause and the victory that was sealed by the Manhattan Project appears to be enduring. The victors of that Cold War, that development of atomic weaponry beyond the Manhattan Project, were all of us on both sides of the Iron Curtain. We gave humanity time to evolve political systems, time for generations to come and go, for hatreds to change. And it gave us time to come to grips with our ability to destroy civilization as we know it. Strategic peace prevailed for five decades. But we must not languish in that victory for victory is that most ephemeral of conditions, fleeting, ennobling but as much a challenge as an accomplishment. I believe in the inherent goodness of humanity and I dislike violence in all of its forms — written, in film, and certainly in reality. But I also recognize that we are a violent species and this is not the first time we have been faced with terrorism.

It also is not the first time that there has only been a single superpower. We have had a number of instances in history where both of those elements prevailed, often in conjunction with one another. We must learn from that experience and we have been singularly ineffective at doing so in the past. The stakes today, however, are a little higher: the persistence of civilization. Would that the people of the world could live in harmony, that reason could prevail over prejudice, that freedom could be recognized in all nations as the right and goal of all people.

Five thousand years of recorded history is a relatively short period of time for the modification of the human race; we are today, as human beings, much as we have been in all of recorded history. We retain those violent impulses of our distant ancestors, drives that may have served them well in their hunt for food, but drives that are ill-suited to the reality of modern weaponry. Nuclear weapons will not go away: the past decade has seen an increase in the number of nations possessing them and number of states and sub-state groups around the world who are actively seeking them. I believe that the role of nuclear weapons in deterring large-scale aggression is a valid and compelling reason for their continuation in our arsenal. They are the big stick that our nation holds that promises the sure defeat of any adversary that dares to offer a strategic challenge to the United States, our friends, or our interests. They do not solve every problem. They do not solve the terrorist problem. They are weapons of *extremis*, I believe, to be thought of only *in extremis*. So in closing, from a historical perspective I would say, let no one forget the horrors of the twentieth century lest they revisit the twenty-first. Let no one forget that with the promise of peace came the prospect of Armageddon. The choice is, always has been, and always will be, ours.

Expertise and Independence: The Role of the Science Advisor

Richard L. Garwin

Philip D. Reed Senior Fellow for Science and Technology at the Council on Foreign Relations; Adjunct Professor of Physics at Columbia University

Well, you've had a good account of the Manhattan Project from the details to the grand sweep. Jerome Karle told us of his separation of plutonium metal in the centrifuge at the University of Chicago. You can read more about it and what happened in the Lillian Hoddeson *et al.* book *Critical Assembly*. With more plutonium, no longer was it necessary to use a centrifuge so that the globules would agglomerate.

My own experience with nuclear weapons began in 1950 when I went to Los Alamos for the first summer as a young faculty member in the Physics Department at the University of Chicago. I was much impressed. I spent the first week or so in the technical library reading the progress reports written every week during the war. This was facilitated by the Oppenheimer policy of non-compartmentalization. So I could read anything that had gone on. It was a good background to making my own contributions to nuclear weaponry, which were in the design and testing of nuclear weapons and soon to the first thermonuclear weapon — the hydrogen bomb.

I worked then on air defense, not at Los Alamos, although I returned every summer for many years. In 1949, of course, the Soviet Union had tested its first nuclear weapon and there was renewed interest in nuclear weaponry in the United States. Few expected the number of U.S. nuclear weapons to peak at 33,000 in 1967 and Soviet nuclear weapons at 45,000 in 1982, each of them many times the power of the Hiroshima and Nagasaki bombs. Our first hydrogen bomb entering into the inventory had an explosive yield of 20 million tons of TNT, a thousand times the yield of the plutonium bomb. And as used to be said in another context, I guess it was by Everett Dirksen, a factor thousand here, a factor thousand there, and pretty soon you're talking about real destructive power. In fact we could destroy not only everybody on Earth but the entire environment with the use of a small fraction of those nuclear weapons.

I had two four-year terms on the President's Science Advisory Committee with Presidents Kennedy and Johnson and then President Nixon. And I chaired several military panels there and have been associated with government service for a long time up to the present: Notably in 1998, as part of the nine-person Rumsfeld Commission to Assess the Ballistic Missile Threat to the United States. So I've seen the strength of our system and many of the weaknesses. I think Maxine Singer is exactly right. We entered World War II totally unprepared in science and technology and in armaments, and very soon, as a result of this total lack of capability, had the flexibility to create anew what was needed. Now we have tremendous capabilities, but we have a bureaucracy which has expanded more rapidly. And with few exceptions, like Steve Younger, there is very little scientific or technical competence at the highest levels of government. So the people who make the decisions in the White House, in the Defense Department, really have to go down a long ways before they find somebody who understands the problems and the potential solutions.

Cindy Kelly mentioned the book I published six months ago with Georges Charpak, *Megawatts and Megatons*. This is a description of everything you need to know about nuclear weapons and nuclear power. It turns out that nuclear weapons are a lot simpler than nuclear power, so we describe them first. They are also a lot simpler to make, unfortunately, and can have much more impact on society. Chapter 10 of our book is called "Making Best Use of Scientists" and addresses this question of contributions of individuals as advisors to people in power versus committees of various types. We go into detail over the President's Science Advisory Committee 1956–1973. I was either a consultant or a member for that entire time. It met every month for two days, 18 people. It had perhaps a dozen panels many of which met two days every month, as well. Half of the panelists were members of the Committee and half were leading experts from around the country. As for the panels, there was no vetting by the White House personnel office which has a hand it seems in every (even temporary) appointment these days — not only in the Bush Administration but progressively worse as time went on in the Clinton Administration and in the Administrations before that.

There are external advisory groups of people who are concerned about the world. I already mentioned Pugwash in connection with Vicky Weisskopf. I've been a council member of Pugwash. There was the Doty Group, the brief name given for Paul Doty, Professor of Biochemistry at Harvard University, who with courage in this country, and with even more courage on the other side in the Soviet Union, began serious contacts with Soviet scientists on arms control and disarmament measures. These helped to lead to the cooperative control of some of the threats and were a background even to the Nunn–Lugar cooperative threat reduction effort of the 1990s.

Another type is the effort of the National Academy of Sciences, Engineering, Institute of Medicine — the National Academies these days. The NAS has a Committee on International Security and Arms Control (CISAC) formed in 1980 with the purpose of meeting twice a year with Soviet scientists which, in my opinion, has had a substantial effect. In part this was due to the luck of having dealt with people who became Gorbachev's science advisors for some years before Mr. Gorbachev came on the scene. And we discuss in Chapter 10 a number of other mechanisms. For instance, there were study groups of the American Physical Society: one in the 1970s on light water reactor safety; in the 1980s on directed energy weapons in the Strategic Defense Initiative; and a current one on boost phase intercept for missile defense.

So many of the innovations, contributions, proposals which turn out to be very good or very bad come from individuals. It is hard for committees to be very innovative. But committees can be deliberative, they can give a good responsible view. The problem is always the tension between expertise and independence. The way you get to be an expert in a field is to work on a problem. Once you work on a problem, you're employed by somebody (the government, a national laboratory, industry) and have a real or apparent conflict of interest. On the other hand, people who are independent are often not experts. So what is the best way to make a committee? In our book we judge that the best is to have people who have proved their integrity, continue to prove it in the scientific and technical community, who may not be expert but who can be briefed by experts and by immersion over a period of time can develop the necessary expertise.

What are our current problems aside from the inherent lack of scientific competence and recognition of scientific problems and solutions in the top levels of our Administration? We don't have a President's Science Advisory Committee. We have a President's Council of Advisors on Science and Technology, but it meets rarely; it doesn't have the staff; nor does it have the independence that PSAC possessed.

We have, unlike the Oppenheimer regime at Los Alamos, serious compartmentalization. The Rumsfeld Commission in 1998 was so concerned about this when we found that people doing intelligence analyses in the community — in the CIA, DIA — would write papers without knowing of the information that was available at different levels. We have both threats and also responses which are proposed without responsible review; and this compartmentalization is often not a security measure but is a way of protecting power, of protecting programs against interference by others who might reach different views or deduce different facts about the program.

Why are things different now from World War II? Well, then there was a universally-shared recognition of the danger. It is hard for young people now in the 21st Century to understand that in the United States there was serious contemplation that we would lose that war, and that genocide would spread throughout the world. Now members of Congress, staffs, and the Administration really do not recognize the current danger. Since September 11 it is a little more apparent. But still we are involved with the highest priority of re-election, of political pork, of unreviewed additions to budgets which only cut into the strength of the responses that we need to make to terrorism. They detract from the reduction of the threat to our survival posed by the 18,000 nuclear weapons still in the Russian inventory, and the thousands of tons of highly-enriched uranium weapon-usable material which is not very well secured over there.

I have mentioned the problems and it is not clear what the solution to them is. We hope to make progress here by meetings like this. The National Academies have a counterterrorism committee with several panels which will report by the end of May, and we hope that this will lead to a more sober assessment of the problems that we face in the world. These are not only political problems, not only problems of national and international security. There are problems which are even more fundamental — the preservation of a democratic government which has to be both effective and free.

When I was on the President's Science Advisory Committee there were many policy problems. There were difficulties of advising the President and yet being able to testify to the Congress. We were told by President Nixon at one time that we should feel free to give Congress the benefit of our expertise. But somehow I don't believe that he really meant it, because free testimony to Congress was met by retribution from the White House — even to the dissolution of the President's Science Advisory Committee. But we discussed such things before testifying and I quoted the code of ethics for government employees. It would provoke laughter if you read it now, because it says that the government employee has no private word — there is no way you can promise somebody something — you have a public responsibility. I published many of these things in various journals (you can find some of them on my website at www.fas.org/rlg). I think that we really do need to go back to the fundamentals which have created the strength of our government and the strength of our scientific community.

These essentials are independence and freedom and recognition of reality. They are being overwhelmed by concerns that are more urgent but far less important.

The Future of Nuclear Deterrence

Richard Rhodes
Independent journalist and author

I said this morning that I would talk about how one might have nuclear abolition, so I will be either a devil's advocate or perhaps an angel's advocate here. It is appropriate that Steve Younger, as someone who is professionally concerned with our national security, believe that nuclear weapons will not go away. But I would point out that it's ultimately not nuclear weapons that deter. If it were then September 11th would not have happened. It is ultimately the knowledge of how to make nuclear weapons that in the long run is the deterrent. There is some slim evidence in the record that we were deterred against using a nuclear weapon on Germany during the Second World War — provided of course we would have had one — by Groves's concern that if it were a dud, and landed on German territory, the German scientists could have recovered the uranium or plutonium and finished their bomb. That's deterrence at the level of knowledge. I think it's clear — and Steve's use of the point about the abrupt decline in man-made death after 1945 — I think it's clear that nuclear weapons deterred the two superpowers from going to war. If that's an experiment, there's the evidence. But we're not in that situation anymore. Today we're in a situation of radical asymmetry between the conventional forces the United States is capable of deploying and the conventional forces that other entities — any other entities, basically — are capable of deploying. We had evidence of what that asymmetry means on September 11th. It's remarkable not so much that the World Trade Center was attacked, but that the Pentagon was attacked successfully.

There's another reason for thinking about how to reduce and eventually eliminate nuclear weapons, and that is something that Richard Butler, the Australian ambassador who was head of UNSCOM, speaks about quite frequently: the fundamental asymmetry in a legal sense and a moral sense in our telling other nations that nuclear weapons are vital to our security but that they're not vital to their security and they should not build them. So if you think of abolition in terms of the time it takes to deliver a weapon to target, you can imagine a world where there might be no physical nuclear weapons but only the carefully maintained knowledge of how to build them. In the early years of the Cold War, when Curtis LeMay's B-29s and then B-52s were flying their missions around the world, delivery time from base to target was perhaps what, 20 hours? Today it is 30 minutes from an ICBM silo, perhaps 15 minutes or less from a nuclear submarine. So if eventually nuclear weapons were taken apart and put on the shelf in separate pieces then how long would the delivery time be then? If nuclear weapons were actually broken up, and their materials diluted and so forth, and only the moth-balled factories were maintained and the laboratories with the knowledge of their design and operation was maintained, what would the delivery time be? Would it be three months? The catch of course is the possibility of someone cheating. The Acheson–Lilienthal plan discussed this question in 1946, at a time when this whole era was just beginning, and imagined that it would be possible to control the entire nuclear enterprise from the mine head to the finished nuclear power plants if they were then distributed in a relatively equitable way around the world (the nuclear power plants and the laboratories where the benevolent effects of nuclear energy were being developed). Then, if someone started to cheat and was detected, said the Acheson–Lilienthal plan, it would be possible for there not to be an army, as Baruch kept looking for in this plan. He asked, "Where's your enforcement regime?" And Oppenheimer answered that it's built into the system, because if one nation starts to cheat then the other nations can immediately begin to develop their arsenals, so that one nation starting clandestinely to work on nuclear weapons would be an act of war. Baruch really didn't get the idea, it was not clear to him. Whether it would have passed muster in the U.N. is yet another question. But this assumes that you detect the cheating. It seems to me if there's a question here about what lessons are to be learned from the Manhattan Project, the fundamental lesson is that patriotism and loyalty can be evoked and marshaled to protect and defend our nation.

There's a great, clear need for improved science and technology to make the world more transparent, not only to clandestine nuclear activities, but also to a more difficult problem indeed, to clandestine biological and chemical activities. Nuclear weapons will not go away. But we didn't quite feel that way about biological weapons. Our nation decided that the way to go where biological weapons are concerned was to agree not to make them anymore, and to encourage other nations by our example not to proliferate biological weapons. So if there's an important scientific and technical task for our nation today I would argue that simple equity as well as the radical asymmetry of present-day military systems of the world point strongly in the direction of reductions of nuclear weapons, elimination of nuclear weapons, while maintaining the scientific and technological skills necessary to follow them. That same effort towards developing more transparency, more technologies of transparency as well as more regimes and agreements and treaties related to transparency, has obvious collateral immediate benefit in the war against terrorism, so what's to lose? There's a physical, chemical equivalent if you will, to the understanding in the biology community today: that efforts towards dealing with bioterrorism are basically indivisible from efforts to improve the public health systems of the nation and the world, and I think that analogy applies to the nuclear side as well. So there's a task, and an immensely valuable task.

Chapter 7: Closing Reflections

Reflections on the Manhattan Project: Consequences and Repercussions

Dr. James Schlesinger
Counselor to the Center for Strategic and International Studies (CSIS); Senior Advisor to Lehman Brothers; Former Secretary of the Department of Energy; Secretary of Defense; Director of the Central Intelligence Agency; and Chairman of the Atomic Energy Commission

Thank you, Cindy. We all owe a debt of appreciation to Cindy Kelly, who has been the spark plug behind establishing the Atomic Heritage Foundation, as well as the spark plug in organizing this conference — intended to memorialize the Manhattan Engineering District, and the Manhattan Project, while memory lasts. Thank you Cindy!

I'm to talk a bit about some reflections on the Manhattan Project and its repercussions. In strategic, as well as scientific terms, the development of the atomic bomb was arguably the single most dramatic event of the last century. As far as the public was concerned, it was carried on in almost total secrecy. Britain and Canada knew about it. Also the Soviet Union knew about it — through more informal measures (laughter). At the Potsdam Conference, President Truman confided to General Secretary Josef Stalin about the existence of the bomb; yet Stalin seemed to be indifferent. The thought then was that Stalin did not grasp the consequences or the implications. The reality was he already knew — and was trying to downplay its significance.

Revelation of the Manhattan Engineering District's product occurred over Hiroshima and subsequently Nagasaki in the summer of 1945, as American troops in the Philippines were preparing to invade the Japanese home islands. Its impact was immediate and immense, both strategic and diplomatic — and continues to unfold today. What had occurred was, to use the traditional phrase — unleashing "the Fires of Prometheus."

A question implied in the last panel. Should we have desired to unleash those fires? That question was never raised at the time because of the fear that the Germans would get there first. As Harry Truman triumphantly acclaimed, "We won the race with the Germans." Still should we be pleased with the development of the Bomb? The answer is both yes and no.

Let me talk about the immediate consequences. The first reaction was simple disbelief. (That was reflected, of course, in that theory that Stalin did not understand the bomb's significance.) To illustrate that disbelief — a little history: Early in 1945, Lieutenant Commander Richard King, later an admiral, was dispatched from the War Department with an eyes-only message intended for General Curtis LeMay, who was directing the air war against Japan. The message ordered LeMay to construct an airbase on Tinian Island, and to give it the highest military priority. General LeMay read the message, turned to Lieutenant Commander King and said (rather characteristically), "Sonny, don't your bosses in Washington know that we've got a war going on out here?"

Let me also tell you about Harold Agnew's story. Harold Agnew had been one of Fermi's students. During the war, his parents kept getting letters. "What's wrong with Harold? Why isn't he part of the war?" He went to work at Los Alamos, and, because of those letters, he kept begging to get directly involved in the war. As a result, he was on that companion plane that overflew Hiroshima. As soon as the weapons had been dropped (despite some failure to understand just why Japan had surrendered), Tinian had suddenly become the leading tourist attraction in the Western Pacific. Scores of tourists, mostly military, would visit the base from which the Enola Gay had flown. Luis Alvarez, who had taken Harold out to the Pacific, assigned to him the humble duty of serving as tour guide for those tourists — for which he had some notable stage props. There was the container in which Fat Man had arrived. And there was the container in which the plutonium core for the Nagasaki weapon had come. Anyway, one of the parties of tourists included a four-star navy admiral by the name of Welch. He listened to Harold's narrative about what had happened, and responded, "Just a minute, young man. Are you saying that a weapon like that is the equivalent of thousands of tons of TNT?" Harold meekly replied, "Yes, sir." The admiral responded, "You may believe that, but I don't have to." With that he stood up and left, and the entire party followed him out.

The human mind finds it a little hard to deal with scale changes. These stories are among the many which reflect the widespread disbelief at the time. An immediate consequence of the development of the atomic bomb was to bring World War II to an abrupt end. The surrender of Japan reflected the decision by the emperor to overrule his military commanders who were prepared to fight to the bitter end. Truman's decision has been much argued about. I had friends who were on Luzon in August of 1945. They were awaiting the order to invade the Japanese home islands. One should read the wartime intercepts recently published by the Central Intelligence Agency. Apparently, the Japanese had anticipated that Kyushu, the southernmost of the main islands, would be the invasion site — and were steadily adding to the divisions that were already there. It would have been slaughter. You will understand that my friends, who were poised for the invasion, never had any doubt that Truman had made the right decision.

Countless lives were saved by Truman's decision. Not merely the lives of American troops, but Japanese troops as well. Well, what followed was some feeling of guilt. Much of the effort to develop the bomb had been directed against Germany — on the premise that Heisenberg and his colleagues were competing with us and might get there first. After the German surrender, concern about bombing the unsuspecting Japanese began to spread through Los Alamos. Science would like its accomplishments, its blessings to be wholly uncontaminated. Just as Alfred Nobel was earlier horrified by the use of TNT in war — and therefore established the Nobel Peace Prize.

Further consequences. We have an open society. The desire was to get as much information as possible out into the public domain. That resulted in the *Smythe Report* — and subsequent releases of information. This was wholly consistent with the ethos of Science that you tell as much as you possibly can. Science should not be narrowly nationalistic, but international, and it should be marked by a steady exchange of scientific information. One notable result of that attitude in the period after 1945 is that we declassified the technology of magnetic separation, which we ourselves had abandoned. To say the least, that was rather ethnocentric on our part. We declassified that information because we decided that the use of magnetic separation was not cost-effective for the United States. Saddam Hussein thought otherwise. As a result, in 1991, he came very close to acquiring nuclear weapons through using the magnetic separation technology that we had earlier declassified. Thus, 45 years later, the release of that information came back to haunt us.

What about some of the longer-run repercussions? Earlier I asked whether we really wanted to unleash the Fires of Prometheus? And my answer was both yes and no. Throughout the Cold War, the existence of nuclear weapons clearly was beneficial, and most notably during the early years of the Cold War, when France and Italy seemed to be teetering. Those early postwar years was a time characterized by Barbara Ward of *The Economist* in her book *The West at Bay*. Today, by contrast, with ever-growing fears regarding proliferation, the utility of nuclear weapons seems increasingly questionable. That is clearly reflected in President Bush's criticisms about Iraq, and his concern about what he called "The Axis of Evil"— that these nations might acquire nuclear weapons. So today it is far less clear.

Nonetheless, it was quite clear in the Cold War. Let me review the geopolitical and the strategic significance of those weapons at that time. They may well have been the salvation of Western Europe. At Lisbon in 1952, the NATO allies had agreed to put together a major conventional capability — some 90 divisions — to match the forces of the Soviet Union and its allies in the Warsaw Pact. The Lisbon goal of 90 divisions was never achieved. We failed to put together that conventional force that would have served as a deterrent to what was then almost overwhelmingly a conventional capability of the Warsaw Pact. Instead, we relied on nuclear weapons to neutralize the Soviet/Warsaw Pact threat. To be sure, it is not clear that the Soviet Union ever intended to move against the West, but the existence of its military capabilities were consistently used for political pressures against individual European countries, most notably those on the periphery of the Warsaw Pact.

One notable irony is that the greatest misgivings about nuclear weapons occurred in Europe, and were reflected in the anti-nuclear movements of the 60's, 70's, and 80's. The existence of nuclear weapons in the 50's and 60's permitted the Western world collectively (and even the United States under the New Look) to deter that presumed threat, while continuing to reduce our forces.

In later years, as the Soviet Union developed a counter-deterrent, nuclear deterrence became ever more difficult. Charles de Gaulle — to justify the *Force de Frappe*, kept raising the question: Would the Americans who held the umbrella over Western Europe be prepared to sacrifice New York for Hamburg? It was a logical question to raise, as Soviet strategic capabilities grew, but it was rather theoretical. The subjective judgments of the Europeans and of the Soviets were starkly different. The Europeans wanted 100% certainty that the Soviet Union would be deterred. Anything that might weaken deterrence caused alarm. By contrast, the Soviets, who believed that history was on their side, would be deterred by even a low probability, say 10%, of a nuclear response. *Any* significant probability of a nuclear response was just too risky. It is interesting, moreover, that it was that massive Soviet military effort, amounting to some 40% of total Soviet economic output, that ultimately caused the collapse of the Soviet Union in 1991.

Today, with the bilateral confrontation of the Cold War ended, and the spread of weapons an increasing threat, the result of unleashing those Fires of Prometheus looks far more problematic — and we might today wish that we never had opened that Pandora's box, as Steve Younger mentioned in his remarks.

Yet, one should also bear in mind that the concern about the possible spread of nuclear weapons traces right back to 1945. We have done far better than we ever imagined in the decades following World War II in slowing down, if not preventing, proliferation. President Kennedy thought and stated that within a decade, 15 nations or so might have nuclear weapons. One revealing comment was made by one of our leading scientists — I think it was Leo Szilard — "What happens when Swaziland gets the weapon?" But those weapons have been slower in spreading than was thought. C. P. Snow, who wrote interestingly about the two cultures, said that within a decade, 15 or 20 nations will have the bomb, and within another decade, some of those weapons would be used in anger.

This is not to provide comfort to you in this changed era, but it does indicate that we have done far better in controlling the spread of nuclear weapons than we anticipated. To be sure, that partly reflected the strong disciplines of the Cold War itself, which have now broken down.

Finally, some words about civilian nuclear power. The eagerness to develop civilian nuclear power was driven, perhaps in large degree, by post-Hiroshima guilt. There was a widespread feeling that we had to demonstrate that the atom could be economically useful rather than merely destructive. The Joint Committee on Atomic Energy accelerated the development of nuclear power. Indeed, the compressed history of nuclear power development reflected that continuous congressional prodding. It was a remarkable record of compression. As I observed when I was chairman of the AEC, it was as if one took the entire history of aviation from Kitty Hawk to the development of the 747 and squeezed it down to a decade and a half.

In 1954, President Eisenhower announced the Atoms for Peace program, which was the basis for the International Atomic Energy Agency. Premised on the hope that the atom could be limited to peaceful purposes, it rested on a proposed "deal" in which most nations, excluding the few nuclear weapons states, would not seek to develop nuclear weapons — and in exchange would be given scientific and technical information on the atom. Both the premise and the "deal" were widely derided as naïve. Nonetheless, I do note that that premise of the peaceful atom is what today drives our policy of deploying large nuclear reactors in North Korea. I also remind you that North Korea is a member of "The Axis of Evil."

In those early years, nuclear power appeared to be an easy solution to our energy problem. You'll remember Louis Straus, the Chairman of the Atomic Energy Commission, opining that we would produce electric power that was too cheap to meter. The Arab oil embargo of 1973 led to far greater emphasis on nuclear power. It was reflected in President Nixon's Project Independence. That stress on nuclear power as a means of reducing our oil dependency was subsequently embraced by President Ford. The utility industry responded, building a number of plants and planning a large number of others. And then Three Mile Island came along — and subsequently Chernobyl. And that, at least for the time being, has derailed nuclear power. In my judgment, nuclear power will come back someday — but not very soon. For the time being, the economics are against it. Combined-cycle, gas-fired capacity is far cheaper to build and to operate.

In that regard, the time and cost of nuclear construction highlights a painful irony, as one looks back on the Manhattan Engineering District.

In 1943, we scarcely knew what a nuclear reactor might look like. Then at Hanford, we constructed a production reactor in something like 15 months. Today the planning process for a reactor alone vastly exceeds 15 months.

Before I close, I should mention two other consequences. As Maxine Singer has indicated, as a result of the Manhattan Project there was an acceleration of government support for science — both for defense and for civilian purposes. Science entered into what has been a golden era. That was in sharp contrast to World War I, when the American Chemical Society offered its services to the War Department. It was turned down with the response that we already had a chemist working for the War Department. And finally, there is nuclear propulsion. Driven by Admiral Hyman Rickover, it was nuclear propulsion that permitted the deployment of the Polaris, Poseidon, and Trident submarines that provided an almost invulnerable deterrent — one that could not be destroyed by another country. That had wisely been foreseen by President Eisenhower, who had spoken of "the Polaris submarine weapon-system that could silently hide in the depths of the sea."

Let me conclude: the development of the atomic bomb during World War II by the Manhattan Engineering District was a dramatic episode, perhaps the most dramatic event in the 20th century. It has had massive repercussions — with consequences both intended and unintended. Some of those were foreseen, and some that were "foreseen" have never materialized. We did not have the Doomsday that some projected during the Cold War. That reflected, I believe, the enormous caution of at least that generation of political leadership. Regrettably, that does not imply that all political leaders will be similarly cautious in the future. Nor, as I have indicated, has proliferation occurred at the pace that was once projected. Clearly, though, it has now become the world's most pressing problem.

One may hope, of course, that scientific developments bring blessings that are uncontaminated. Yet, good and evil, as Steve Younger has already pointed out, will always be with us. The blessings of science will never be unalloyed. It is unrealistic to think so. Like all human endeavors, science is Janus-like and faces in two directions.

Thank you very much.

Appendix A: Program

Atomic Heritage Foundation
Symposium on the Manhattan Project
April 27, 2002
Carnegie Institution of Washington
1530 P Street N.W., Washington, D.C., U.S.A.

9:30 AM Opening Remarks: A History Worth Preserving

Senator Jeff Bingaman (NM)

Dr Everet H. Beckner, Deputy Administrator, NNSA (U.S. DOE)

10:15 AM The Manhattan Project: A Millennial Transformation

Richard Rhodes, Pulitzer-prize winning author of *The Making of the Atomic Bomb*

Stephane Groueff, author of *The Manhattan Project: The Untold Story of the Making of the Atomic Bomb*

11:00 AM The Allies and the Atomic Bomb

The British contribution, the role of Sir James Chadwick, and the war effort. How Vannevar Bush and James Conant transformed government's relationship to industry and academia, with lessons for today.

Andrew Brown, author of *The Neutron and the Bomb* (biography of Sir James Chadwick)

James Hershberg, author of *James B. Conant*

Kai Bird, co-author with Martin Sherwin of a forthcoming book on J. Robert Oppenheimer

12:00 PM Lunch Break

1:15 PM–2:15 PM The Military and Science in the Crucible of War

Tensions between scientists and the military on secrecy and deadline-driven research, and how the experience changed American science and the military. Focus on the roles and relationships of General Groves, Oppenheimer, Teller, Lawrence, and Szilard, with lessons for today.

Robert S. Norris, author of *Racing for the Bomb: General Leslie R. Groves, the Manhattan Project's Indispensable Man*

Gregg Herken, author of the forthcoming book *Brotherhood of the Bomb:*

A Tale of Science, Power and Loyalty (about Oppenheimer, Lawrence, and Teller)

William Lanouette, author of *Genius in the Shadows* (biography of Leo Szilard)

2:15 PM–3:15 PM Speaking from Experience

Manhattan Project veterans contribute their first-hand perspectives at the Symposium.

Benjamin Bederson, Special Engineer Detachment at Los Alamos

Jerome and Isabella Karle, Metallurgical Laboratory at the University of Chicago

Maurice Shapiro, Group Leader, Ordnance Division at Los Alamos

Other Manhattan Project veterans who plan to attend and may contribute from the audience are Philip Abelson, Gerhart Friedlander, William Golden, Arnold Kramish, Ralph Lapp, and Ernest Tremmel.

3:15 PM–3:30 PM Break

3:30 PM–4:30 PM Lessons of the Manhattan Project for the 21st Century

Experts reflect upon the factors that were critical to the Manhattan Project's success, particularly how the government was able to marshal the resources of industry and academia. The panel discusses the lessons for meeting today's national security challenges.

Maxine Singer, President of the Carnegie Institution of Washington

Stephen Younger, Director, Defense Threat Reduction Agency

Richard Garwin, Senior Fellow for Science and Technology, Council on Foreign Relations, New York, and Adjunct Professor of Physics at Columbia University

Richard Rhodes, author of *The Making of the Atomic Bomb*

4:30 PM–5:00 PM Closing Reflections

Honorable James R. Schlesinger, Senior Advisor Lehman Brothers, and former Secretary of Defense and of Energy

Appendix B: Participants

Benjamin Bederson is Professor of Physics Emeritus at New York University and Editor-in-Chief Emeritus, American Physical Society, editor of *Advances in Atomic Molecular and Optical Physics* (Morgan Kaufmann, 1997), and author of "SEDs in Los Alamos: A Personal Memoir" (*Physics in Perspective*, Vol. 3, 2001).

Everet Beckner is Deputy Administrator for Defense Programs at the Department of Energy's National Nuclear Security Administration (NNSA), responsible for the nation's nuclear weapons complex. He recently retired as Vice President at Lockheed Martin and previously served as the Energy Department's Principal Deputy Assistant Secretary for Defense Programs (1991–1995). He has also worked at the Sandia National Laboratories and has a Ph.D. in physics.

Senator Jeff Bingaman graduated from Harvard University in 1965 and earned a law degree from Stanford University in 1968. After a year as New Mexico Assistant Attorney General and nine years in private law practice, he was elected Attorney General of New Mexico in 1978 and to the U.S. Senate in 1982.

Kai Bird is a Fellow at the Woodrow Wilson International Center for Scholars. He is the biographer of John J. McCloy and McGeorge and William Bundy and co-editor of *Hiroshima's Shadow: Writings on the Denial of History and the Smithsonian Controversy*. He is co-author with Martin Sherwin of a forthcoming book on J. Robert Oppenheimer.

Andrew Brown trained and worked as a physician in London before moving to the United States in 1990. He has been practicing as a radiation oncologist in New Hampshire since then. Author of *The Neutron and the Bomb*, a biography of Sir James Chadwick, he is currently writing a biography of Desmond Bernal, the X-ray crystallographer and radical.

Richard Garwin is Philip D. Reed Senior Fellow for Science and Technology at the Council on Foreign Relations and an Adjunct Professor of Physics at Columbia University. He is a member of the National Academy of Engineering and of the Institute of Medicine. In 1996, he received the Enrico Fermi Award.

Stephane Groueff is a Bulgarian-born journalist who fled Bulgaria in 1944, worked for *Paris Match* and became its New York Bureau Chief in 1958, serving until 1978. He is author of *Manhattan Project: The Untold Story of the Making of the Atomic Bomb* (Little Brown-Boston, 1967), *Crown of Thorns* (1998), and a recent autobiography.

Gregg Herken is a historian and the Curator of Military Space History at the Smithsonian Institution's National Air and Space Museum in Washington, D.C. He is the author of three books on nuclear history and the biography, *Brotherhood of the Bomb: The Tangled Lives and Loyalties of Robert Oppenheimer, Ernest Lawrence, and Edward Teller* (Henry Holt and Company).

James G. Hershberg is Associate Professor of History and International Affairs at the George Washington University, with expertise in the international history of the Cold War, nuclear history, and contemporary international relations. He is the author of *James B. Conant: Harvard to Hiroshima and the Making of the Nuclear Age* (Knopf, 1993; Stanford, 1995).

Dr. Isabella Karle has made fundamental contributions to the theory and practice of electron diffraction of vapors and X-ray diffraction of crystals. She is a senior scientist at the Naval Research Laboratory where she has been since 1946. She has received the National Medal of Science and the Bower Award from the Franklin Institute.

Dr. Jerome Karle's scientific work has mainly concerned various areas of structural chemistry and the development of useful analytical techniques from physical theory by mathematical means. A Nobel laureate, Dr. Karle is currently Chief Scientist for the Structure of Matter at the Naval Research Laboratory, Washington, D.C.

Cynthia C. Kelly is President of the Atomic Heritage Foundation, dedicated to the preservation of the history of the Manhattan Project and the Atomic Age. For over twenty years, she was a senior manager at the Environmental Protection Agency and Department of Energy and received a Distinguished Career Service Award in 1999.

William Lanouette has written extensively on the politics of nuclear weapons and nuclear power. He was Washington Correspondent for *The Bulletin of the Atomic Scientists* before joining the U.S. General Accounting Office in 1991 as a Senior Analyst for energy and science issues. He wrote *Genius in the Shadows: A Biography of Leo Szilard, the Man Behind the Bomb* (Scribners, 1992; University of Chicago Press, 1994).

Robert S. Norris has been a research associate for almost twenty years at the Natural Resources Defense Council in Washington, D.C, covering nuclear weapons issues. As an author of the multi-volume *Nuclear Weapons Databook* series, and of numerous articles, he has written extensively about the nuclear programs of the United States, Soviet Union/Russia, Britain, France, and China. He is author of *Racing for the Bomb: General Leslie R. Groves, the Manhattan Project's Indispensable Man* (Steerforth Press, 2002).

Richard Rhodes is the author of 19 books including *The Making of the Atomic Bomb*, which won a Pulitzer Prize in Nonfiction, and *Dark Sun*, one of three finalists for a Pulitzer Prize in History, that continued the story of nuclear weapons development in the early Cold War years. Rhodes has written extensively about nuclear issues and lectured widely in the United States and abroad.

James R. Schlesinger is counselor to the Center for Strategic and International Studies (CSIS) and Senior Advisor to Lehman Brothers. He has had a very distinguished career, including serving as the first Secretary of the Department of Energy (1977–1979), Secretary of Defense (1973–1975), Director of the Central Intelligence Agency (1971–1973), and Chairman of the Atomic Energy Commission (1971–1973).

Maurice M. Shapiro is Visiting Professor, University of Maryland. After the Manhattan Project, he had a distinguished career in the field of cosmic rays and neutrino astrophysics. Dr. Shapiro is director of the International School of Cosmic Ray Astrophysics that holds biennial courses for graduate students and young researchers in Erice, Italy.

Martin J. Sherwin, a professor of history at Tufts University, is the author of the prize-winning *A World Destroyed: The Atomic Bomb and the Grand Alliance*. He is co-author with Kai Bird of a forthcoming book on J. Robert Oppenheimer.

Maxine Singer has been President of the Carnegie Institution of Washington since 1988 after serving as chief, Laboratory of Biochemistry, for the National Cancer Institute, and making many contributions in the fields of biochemistry and molecular biology. A recipient of the National Medal of Science, she is a vocal leader in science policy and ethical issues.

Stephen Younger is the Director of the Defense Threat Reduction Agency, Department of Defense, responsible for most of the cooperative threat reduction work being carried out with states of the Former Soviet Union. With a doctorate in theoretical physics, he was the associate laboratory director for nuclear weapons at Los Alamos Laboratory, Los Alamos, NM, and is a leading expert on nuclear weapons technology.

PART II: A PLAN FOR PRESERVING THE MANHATTAN PROJECT

Interim Report to Congress

Preserving America: A Strategy for the Manhattan Project

Note to the reader: The following is a *draft* of an interim Report to Congress that presents some immediate priority actions for preserving the history of the Manhattan Project.[42] In order to fulfill the requirement in the 2002 Appropriations Act (H. R. 107–258), the Department of Energy awarded a grant to the Atomic Heritage Foundation "to develop a plan for preserving the history of the Manhattan Project." This report provides recommendations for what Manhattan Project properties should be preserved as well as steps that should be taken to capture other aspects of the Manhattan Project history. Key considerations for determining which properties to preserve are their national significance, how well the properties represent one of the themes of the Manhattan Project, the integrity of the properties, feasibility and costs of restoration, and the costs of long-term maintenance and operation. In addition, Congress requested that the report identify management alternatives including one or more agencies that could serve as the long-term steward for these properties.

The Report presents two alternative five-year plans for preserving significant properties, equipment and artifacts, and interpreting this history for the American public. The first option is the **Essential Manhattan Project Plan** and costs $48 million over the five years; the second option is the **Enriched Manhattan Project Plan** and costs $74 million over five years. We welcome your comments on this draft.

Atomic Heritage Foundation

Cynthia C. Kelly, President
Atomic Heritage Foundation
910 17th Street, NW, Suite 408
Washington, DC 20006, USA
Phone: 202-293-0045
Fax: 202-293-0699
info@atomicheritage.org
www.atomicheritage.org

[42] H. Rpt. 107–258.

Preserving America: A Strategy for the Manhattan Project

The Manhattan Project, the top-secret effort in World War II to make atomic bombs, was one of the most significant undertakings in American and world history. Yet this history has been "largely invisible" to the American public.[43] Few Americans know about the mammoth, $2.2 billion construction project that created the equivalent of the automobile industry and employed 125,000 people who lived in "Secret Cities."

The Manhattan Project indelibly shaped American and world history. As the first nation to possess nuclear weapons, the United States became a military Super Power. While the Cold War brought a nuclear arms race, recognition of the likelihood of "mutually assured destruction" has prevented another full-scale world war for nearly 60 years. The legacy of the Manhattan Project still profoundly influences the course of world affairs as nations grapple with the threat of nuclear weapons in the hands of "rogue nations" and terrorist groups.

Preservation of the properties of the Manhattan Project has been given little attention. Enshrouded in secrecy for nearly 60 years, most of the production facilities and laboratories are located "behind the fence," where only those with security clearances are allowed. Out of public view until recent years, few people were aware that some Manhattan Project properties still existed. Abandoned in the 1950s, the "V Site" buildings stood as lonely sentinels on a remote mesa at Los Alamos for decades. In the mid-1990s, plans were drawn to bulldoze them down along with dozens of others.

With the end of the Cold War, Congress directed the Department of Energy to focus on the demolition of surplus properties and the environmental restoration of the former Manhattan Project and Cold War

[43] Thomas E. Marceau, David W. Harvey, Darby C. Stapp, *et al.*, *History of the Plutonium Production Facilities at the Hanford Site Historic District, 1943–1990* (Battelle Press, 2002), p. 1.1.

sites. In recent years, this cleanup process has been accelerated. The majority of the Manhattan Project properties have already been lost and most of the remaining ones are scheduled for demolition. Decisions must be made soon — before it is too late — on whether to preserve some of the remaining properties for education, interpretation and commemoration of the Manhattan Project.

On March 3, 2003, President George W. Bush reinforced the importance of preserving some of the Manhattan Project properties through Executive Order #13287, "Preserve America." The Executive Order calls upon the Federal government to "provide leadership in preserving America's heritage by actively advancing the protection, enhancement, and contemporary use of the historic properties owned by the Federal government." In addition, the Order directs the Federal government to seek partnerships in order to "promote local economic development and vitality through the use of historic properties in a manner that contributes to the long-term preservation and productive use of those properties."[44]

The recommendations in this interim report respond to the Executive Order by proposing alternative uses for some of the Manhattan Project properties and taking steps to preserve others. At all three major Manhattan Project sites, State and local governments are very interested as they recognize the potential economic returns from heritage tourism.[45] Notwithstanding the interest of state and local governments in preserving these facilities, they may not have the funds available to aid in their preservation, which can be a costly endeavor.

Evaluation of the Manhattan Project Properties

As discussed briefly above, there is no question as to the significance of the Manhattan Project to American and world history. As the Advisory

---

[44] George W. Bush, President, *Executive Order No. 13287 "Preserve America,"* 3 Mar. 2003.

[45] Valentino, Joe, Executive Director, *Oak Ridge Convention and Visitors Bureau* (ORCVB), Mar. 2003. Advertising Oak Ridge as "World War II's Secret City" has contributed to the city witnessing a 25% increase in Hotel and Motel Tax Revenue.

Council on Historic Preservation stated, "It is imperative that our memory and recognition of the Manhattan Project as a watershed in the history of the nation and the world be preserved for future generations."[46]

The Department of Energy has laid considerable groundwork for these recommendations over the past ten years. In compliance with the National Historic Preservation Act (NHPA), the Department of Energy has designated Historic Districts at Oak Ridge[47] and Hanford[48] and identified which Manhattan Project properties are "contributing properties" and are eligible for the National Register of Historic Places. Los Alamos has recently conducted an evaluation of its Manhattan Project properties and determined which are eligible for the National Register.[49] In addition, the Department of Energy designated eight properties from the Manhattan Project as "Signature Properties" that "provide the essential core for successfully interpreting for the American public the Manhattan Project mission of developing an atomic bomb."[50]

The Department of Energy has also articulated themes for its Manhattan Project and Cold War history based on the primary functions involved in creating the world's first atomic bombs.[51] By design, during the Manhattan Project most of these functions were carried out in three distinct geographical areas: Los Alamos, NM; Oak Ridge, TN; Hanford, WA.

---

[46] Advisory Council on Historic Preservation, *Recommendations and Preservation Options for Manhattan Project Signature Facilities at Oak Ridge and Hanford Reservations*, Washington, D.C., Feb. 2001.

[47] Peter Souza, ORNL; Glyn DuVall, Nancy Tinker, DuVall & Associates; and Melissa Hart, ORO; *Cultural Resource Management Plan, Department of Energy Oak Ridge Operations Office, Anderson and Roane Counties Tennessee*, Oak Ridge, TN, Jul. 2001. DOE/ORO-2085.

[48] Thomas E. Marceau, David W. Harvey, Darby C. Stapp, *et al.*, *History of the Plutonium Production Facilities at the Hanford Site Historic District, 1943–1990* (Battelle Press), 2002.

[49] Ellen D. McGehee, *Sentinels of the Atomic Dawn* (Los Alamos National Laboratory, Mar. 2003), p. 47.

[50] F. G. Gosling, *The Signature Properties of the Manhattan Project* (Department of Energy, Feb. 2001), p. 1.

[51] Eight themes are fully presented in the Department of Energy's *Linking Legacies: Connecting the Cold War Nuclear Weapons Production Processes to Their Environmental Consequences* (DOE/EM-0319), Jan. 1997.

For Los Alamos, the overarching theme is Nuclear Weapons Research and Development as the laboratory was primarily devoted to the development and testing of the "Trinity" device (tested on July 16, 1945), the "Little Boy" bomb (dropped over Hiroshima on August 6, 1945), and the "Fat Man" bomb (dropped over Nagasaki on August 9, 1945). Of the 51 Manhattan Project properties remaining at Los Alamos, 44 were found to be eligible for listing on the National Register of Historic Places, and at least 10 of the properties were recommended for preservation by the Los Alamos National Laboratory.[52]

For Oak Ridge, the overarching theme is Isotope Separation or production of enriched uranium. Naturally occurring uranium is over 99 percent uranium-238 (U-238) and less than one percent uranium-235 (U-235). The goal of the enrichment process was to achieve over 80 percent U-235 for use in an atomic bomb. Oak Ridge focused on three processes for uranium enrichment: electromagnetic separation, gaseous diffusion, and liquid thermal diffusion. The thermal diffusion plant and more than half the Y-12 plant (all 864 Alpha calutrons) were shut down a month after Japan surrendered. This was possible because of the success of the K-25 Plant which continued to operate smoothly for decades. The rest of Y-12's U-235 operation shut down at the end of 1946, but a few Beta Calutrons (72 of the 288) have continued in use separating isotopes other than uranium ever since. The Y-12 electromagnetic separation process and K-25 gaseous diffusion plant have both been designated "Signature Facilities" of the Manhattan Project.[53]

Y-12 "Racetrack" at Oak Ridge.

[52] Mc Ellen D. McGehee, *Sentinels of the Atomic Dawn* (Los Alamos National Laboratory, Mar. 2003), p. 1.

[53] F. G. Gosling, *The Signature Properties of the Manhattan Project* (Department of Energy, Feb. 2001), p. 4.

Another significant theme at Oak Ridge is Reactor Operations. A third "Signature Facility," the X-10 Graphite Reactor at Oak Ridge was built in just nine months and began operating in November 1943 as a prototype for the plutonium production reactors at Hanford. It produced the first significant amounts of plutonium that were instrumental in designing the "Fat Man" or plutonium-based implosion bomb. The X-10 chemical separation plant used to extract the plutonium proved the feasibility of the bismuth phosphate process used at Hanford.

In the past, Oak Ridge has considered whether to preserve its Manhattan Project properties on a case-by-case basis. Recently, surveys of the remaining properties in each of the principal areas (Y-12, X-10, and K-25, now called the East Tennessee Technology Park or ETTP) have been underway and will provide a basis for a more comprehensive and methodical approach in the near future.

At Hanford, the major themes related to the Manhattan Project are Fuel Manufacturing, Reactor Operations, Chemical Separations, and Plutonium Finishing. The main mission of the Hanford site was plutonium production by irradiating uranium fuel rods and then extracting the plutonium for use at the Trinity site and in the "Fat Man" type bombs. The process by which this highly fissionable element would be produced was nothing short of the transmutation of matter, the alchemist's age-old dream of turning lead into gold.

The T-221 Chemical Separation Plant at Hanford.

Over the next 45 years, with six additional plutonium-production reactors, Hanford was at the forefront in the design and operation of nuclear reactors, not only for making fissionable material for bombs, but also for producing power, creating isotopes, and continuing research in nuclear energy. After evaluating 1,100 properties, 527 were found to be contributing properties within the National Register "Hanford Site Manhattan Project and Cold War Era Historic District." A representative sample of 190 buildings was selected for individual documentation, with 30 of these recommended for preservation in place, public education, or economic development. Twelve of these facilities were constructed for the Manhattan Project.[54]

Basis for Recommendations

The following preliminary recommendations are based on the work of the Department of Energy as well as input from State and local governments, the National Park Service, museum and historical societies, and many other stakeholders who participated in public meetings convened by the Atomic Heritage Foundation and numerous other discussions over the last year. Specifically, public meetings were held in Oak Ridge on March 25 and 26, 2003 and in Richland on April 30 and May 1, 2003. A series of meetings were held in Santa Fe and Los Alamos on April 7 to 9, 2003. The agenda, presentations, recommendations, and media reports from those meetings may be found at www.atomicheritage.org

In addition, this report is influenced by the recommendations submitted to the Department of Energy in February 2001 by the Advisory Council on Historic Preservation.[55] The report stressed the need for "creative and innovative approaches" to preservation, reaching outside the Department for solutions and ideas. The Council also recommended:

- Working with the National Park Service to explore the establishment of a National Park Site,
- Partnerships with State and local governments to promote heritage tourism and market historic properties for rehabilitation and reuse,
- The Department of Energy lead other public and private institutions in a partnership to share the costs of restoring and managing the properties,

---

[54] Thomas E. Marceau, David W. Harvey, Darby C. Stapp, *et al.*, *History of the Plutonium Production Facilities at the Hanford Site Historic District, 1943–1990* (Battelle Press, 2002), p. 1.2.

[55] Advisory Council on Historic Preservation, *Recommendations and Preservation Options for Manhattan Project Signature Facilities at Oak Ridge and Hanford Reservations* (Washington, D.C., Feb. 2001).

- Incentives for contractors to meet historic preservation goals, and
- Strategies that consider the Manhattan Project "holistically" and include properties that relate to the history of the community.[56]

These and other recommendations in the Advisory Council on Historic Preservation's report provided an important framework for the development of this plan.

Cross-Cutting Recommendations

1. Special Resource Study for National Park Units

The Department of Energy's policy is that another agency should serve as the long-term steward for its Manhattan Project resources. As part of preparing this report, representatives of the National Park Service participated in several meetings to explore whether the creation of a National Park unit would be a realistic alternative.

For all three major sites (Los Alamos, NM; Oak Ridge, TN; Hanford, WA) the creation of a National Park unit offers a viable alternative for the long-term stewardship of the properties in conjunction with other public and private partners. No national monument or park currently exists that is dedicated to the history of the Manhattan Project. With its mission to preserve America's heritage, the National Park Service could play a central role in protecting and interpreting these important national resources.

A leading recommendation from all three sites was to pursue authorization and appropriations for a Special Resource Study. As a prerequisite to creating a new National Park unit, Congress must authorize a Special Resource Study to determine if a site meets the criteria established by law for inclusion to the National Park System. The study would examine whether each of the Manhattan Project sites would meet

[56] Advisory Council on Historic Preservation, *Recommendations and Preservation Options for Manhattan Project Signature Facilities at Oak Ridge and Hanford Reservations* (Washington, D.C., Feb. 2001).

the National Park Service criteria for national significance, suitability and feasibility for inclusion in the Park System.

2. Oral Histories of Manhattan Project Veterans

Each year more Manhattan Project veterans are dying and the need to interview and record their histories has gone from urgent to critical. What is not captured on tape now will be lost forever. Often more informative than written accounts, oral histories of the men and women involved in the Manhattan Project provide invaluable perspectives of life within the top-secret project.

There is no inventory of existing oral histories of the Manhattan Project; however most are of the leading officials, scientists, and engineers. Some oral histories are not publicly available as they contain classified or sensitive information.[57] Very few interviews exist in the public domain of other groups such as women, members of the Special Engineer Detachment, or African-Americans. With appropriate levels of funding, the oral histories can be captured on archival quality video tape and presented in documentary films for public broadcasting stations, museum exhibits, and educational programs. In addition, the original tapes and transcripts will be available for scholars and historians.

3. Preservation and Storage of Equipment, Artifacts and Documents

As the Department of Energy accelerates its cleanup and dismantling of Manhattan Project facilities, some first-of-a-kind equipment and artifacts have been lost or are in danger of being lost because of the lack of appropriate storage space. Many of these items have been lost due to lack

[57] For example, Stanley Goldberg interviewed several Manhattan Project veterans in the late 1980's. The tapes are in the Smithsonian's collection. Classified information was included in the interviews under the Knowledge Management Initiative in order to preserve technical information.

of funding for radiological surveys. This results in disposal by burial rather than inspections, and decontamination, as necessary. In order to avoid losing irreplaceable objects, once objects have been identified for preservation and until they are ready to be displayed, the Department of Energy should make arrangements for the curation and storage of such objects and documents.

Preservation Strategies for the Manhattan Project: Two Options

The Manhattan Project marshaled the best of science, mathematics, engineering and technology in a race to build the first atomic bombs. This effort is to save some of the most significant historic properties and first-of-a-kind process equipment and develop interpretive exhibits and displays that will help Americans understand this important chapter in the nation's history.

One overarching goal of this initiative is to promote greater literacy by the American public about nuclear science and the history of nuclear weapons development. Given how important nuclear weapons and nonproliferation issues are in world affairs today, it is important that the public have a better grasp of these issues. Another is to provide the foundation for developing heritage tourism to the former Manhattan Project sites. The initiative will provide for economic development opportunities at these sites that have historically been dominated by a single mission.

In most cases, the cost savings of preservation and interpretation over "decontamination and decommissioning" would be significant. The next draft of this report will include further analysis of surveillance and maintenance costs for different projects but these costs also should be less than disposal of the buildings. Over time these costs should be at least partly offset by the revenues from tourism and private and public donations.

Additional options can be created by adding new recommendations or combining pieces of the following options. There are uncertainties that will be addressed in the next year that will help to shape the plan in 2005 and beyond. However, for purposes of discussion, the following two options present the "Essential Manhattan Project" and the "Enriched Manhattan Project" alternatives.

The Essential Manhattan Project (Option A)

What is essential to tell the story of the Manhattan Project? As discussed earlier, the functions of the Manhattan Project were divided geographically between Oak Ridge, TN; Hanford, WA and Los Alamos, NM, with critical supporting research at the University of Chicago's Metallurgical Laboratory, University of California at Berkeley, and Columbia University and at the Trinity Site, where the first atomic bomb was detonated at Alamogordo, NM. In addition, more than one hundred smaller sites across the country were involved in uranium processing activities, scientific and engineering research, production of equipment and supplies, and other operations. However, these sites are beyond the scope of this Report.

While most of the original facilities have long since been demolished, sixty years later we have a unique opportunity to preserve a few of the original Manhattan Project research laboratories and production facilities as well as examples of first-of-a-kind equipment and other artifacts.

The following list of properties comprises the essence of the Manhattan Project, illustrating the key processes involved at Oak Ridge to produce enriched uranium, at Hanford to produce plutonium, at Los Alamos to design and build the world's first three atomic bombs, and at the Trinity Site in Alamogordo, NM, to test the first bomb:

Oak Ridge: Isotope Separation and Reactor Operations

Oak Ridge has three "Signature Facilities" of the Manhattan Project. The X-10 Graphite Reactor and Y-12 Beta 3 Calutrons are both National Historic Landmarks and unique resources for understanding the Manhattan Project. The goal is to preserve them both for public interpretation, illustrating the themes of reactor operations and isotope separation.

The Beta 3 West track has 30 of its original 36 calutrons virtually intact, though they have been modified for separating a wide variety of isotopes of other elements. Modernization of this vital plant may permit public access sometime in the future. Another alternative is to move a few of the calutrons — already slated for cleanup — to another site easily accessible to the public where they can be displayed in their exact configuration. One candidate is to create a Manhattan Project and Cold War science and technology museum at the K-25 site, an option that is going to be studied by an architectural and engineering firm this fall. Modest funds are included in Option A for interpretation and upgrading exhibits at both the Oak Ridge National Laboratory (X-10 site) and the Y-12 site.

The world's largest roofed building when it was built in 1943, the K-25 plant housed 3,000 stages that comprised a mammoth gaseous diffusion process for separating isotopes of uranium. The architectural and engineering study will present suggestions for preserving and interpreting the K-25 building, another of the Signature Facilities of the Manhattan Project. If the K-25 building is razed leaving just the footprint or interpretive structures, the "Roosevelt Cell," one of the 3,000 stages, could be moved to the Manhattan Project and Cold War science and technology museum at the K-25 site or some other location available to the public.

K-25 gaseous diffusion process building.

The K-29 building, a Cold War era gaseous diffusion plant, is one option for the proposed science and technology museum. It would be easily accessible to the public and allow for viewing of the K-25 site. However, the Department of Energy has concerns about the cost of upgrades to the facility and wants the architectural and engineering study to address alternative sites for the museum.

The new museum would complement the current American Museum of Science and Energy (AMSE) in downtown Oak Ridge. First opened in 1949 and originally named the American Museum of Atomic Energy, AMSE is a leader in showcasing the development of peaceful uses for atomic energy at Oak Ridge and across the country under the leadership of the Department of Energy. The new vision is to revitalize the museum through capital improvements in order to make it the preeminent center for Manhattan Project interpretation and education on the East coast. Exhibits will present the overall history of the Manhattan Project and focus on the social and cultural aspects of Oak Ridge's experience.

The newly proposed museum at the K-25 site would focus on the technical aspects of the Manhattan Project. Exhibits with examples of the original equipment would help the public better understand the science and engineering involved in the three parallel processes used to separate isotopes during the Manhattan Project. Other exhibits will address the issues involved in reactor operations, based on the X-10 Graphite Reactor that served as a prototype for the Hanford reactors. For more details, please refer to the White Paper prepared by William J. Wilcox, Jr., *Preserving Oak Ridge's Historic Sites for Telling the Story of Its Role in the Manhattan Project*, a report produced in response to recommendations from the March 26–27, 2003 Workshop in Oak Ridge sponsored by the Atomic Heritage Foundation.

One of the most significant Manhattan Project buildings at Oak Ridge that illustrates its social and community history is the Alexander Inn, which served as the guest house built to accommodate distinguished visitors such as General Leslie Groves, Enrico Fermi, and Ernest Lawrence. The proposal provides funding to purchase and restore the facility as a welcome center for future visitors to Oak Ridge, to be managed by the Oak Ridge Visitor and Convention Bureau. Funds will allow the restoration and interpretation of the guest rooms where dignitaries stayed and conducted their urgent business of the Manhattan Project.

Hanford: Plutonium Production

The Hanford site was dedicated to the production of plutonium and its properties uniquely represent themes related to Fuel Manufacturing, Reactor Operations (for plutonium production), Chemical Separations, and storage. Two of the icons of Hanford's Manhattan Project history are proposed to be preserved, the B Reactor and T Plant. The B Reactor has been largely restored and the funds requested are to finish this work and provide for exhibits and interpretive materials.[58]

Repairs to the four miles of road, old Route 6, leading from the northwest corner of the Hanford reservation to the B Reactor will allow the public traveling by car or bus to have easy access to the B Reactor site, without having to traverse a large portion of the Hanford Reservation.[59] The roadway will be fenced on either side by barbed wire for security.

Nicknamed the "Queen Mary" because of its huge size at 875 feet long, 85 feet wide, and 102 feet high, the T Plant was built by DuPont as a chemical separation facility to extract plutonium from the irradiated fuel rods. The Department of Energy is currently using the facility as a low-level waste storage facility. Funds are requested to provide for the building's exterior integrity, repairing the roof and other structural work as necessary. Because of contamination within the interior processing areas, under this option the public will view this facility from a distance but learn of its unique design and use of innovative remote control operations by a documentary film.

Two remaining properties from the pre-World War II settlements are the Hanford High School and the First Bank of White Bluffs. A modest funding requirement is necessary to stabilize these properties and preserve for future generations something of the history of those settlers who gave up their homes for the war effort.

---

[58] In 2001, a regulatory document (a ten-year Interim Engineering Evaluation and Cost Analysis — DOE/RL-2001-09) was prepared which identified requirements for allowing limited public access along an established tour route. Room by room hazard mitigation (i.e. lead paint abatement, asbestos encapsulation, tripping hazard mitigation, etc.), facility-wide improvements (electrical upgrades, ventilation system improvements, etc.) and structural studies were mandated. These improvements are almost complete, with the exception of roof replacement and structural enhancements. Routine surveillance and maintenance of the facility is ongoing to protect workers, visitors and the environment.

[59] Plans for repaving the proposed access road must be coordinated with the ongoing soil remediation in the B/C area. The expected completion date for the soil remediation is FY 2006.

Los Alamos: Designing, Building and Testing the Bomb

The essential Manhattan Project properties at Los Alamos are three properties which uniquely represent the work done on the three bombs designed and built at Los Alamos during the Manhattan Project. The first one, called the "Trinity device" or "Gadget," was associated with the "V Site." The "High Bay Building," located at the "V Site" was where the implosion lens were assembled and tested for the plutonium-based bomb. Because this site received a Save America's Treasures grant, only modest additional funds are requested. For the foreseeable future, public visitation will be restricted because of its location behind two security perimeters. Moderate funds are also requested for the restoration and interpretive story board for the "Concrete Bowl" nearby.

The second site is the "Gun Site," where "Little Boy," the uranium-based bomb dropped on Hiroshima was tested. This site will be partly restored also using Save America's Treasures grant funds. The main building will serve as an interpretive center for the Manhattan Project with regular public access. Funds are needed for interpretive exhibits and displays.

The "Gun Site."

The third site is the Quonset hut where the "Fat Man," or plutonium-based bomb dropped over Nagasaki, was tested and assembled. The three sites together with the "Trinity Site" (see below) tell the story of the design, assembly, and testing of the world's first three nuclear devices.

Funds are also requested to restore and interpret the Master Cottage built by the Los Alamos Boys Ranch School where J. Robert Oppenheimer and his family lived during the Manhattan Project. The Los Alamos Historical Society will have title to the property, which could be interpreted by the National Park Service.

Finally, funds are requested to enable the Los Alamos Historical Society to improve the storage of its archival records. The current conditions present a potential fire hazard and inhibit gifts of additional documents.[60]

The Trinity Site

The Trinity Site at Los Alamos.

The world's first atomic bomb, "Gadget," was tested on July 16, 1945 at Alamogordo, NM at what is known as the "Trinity Site." Funds are requested to restore the historic facilities that are now part of the White Sands Missile Range managed by the Department of Defense. The goal is to provide for more opportunity for public visitation and, under Option B, to save some of the original Manhattan Project structures that have never been restored. Under this option, funds are requested to repair the original McDonald Ranch house and upgrade its exhibits. The last restoration work was completed in 1984. Finally, funds are requested to capture oral histories and produce a short documentary film of the Manhattan Project veterans who were involved in and witness to the world's first atomic explosion.

The Enriched Manhattan Project (Option B)

This option enhances the preservation strategy by including important additional properties associated with major themes of the production processes at Hanford, and the communities of Oak Ridge and Hanford. It also provides for additional investment in exhibits and interpretation of the Manhattan Project history at visitor centers at Los Alamos and Richland, which is currently being planned in conjunction with other

[60] The Los Alamos Historical Society's Manhattan Project Archives are stored on the third floor of Fuller Lodge in "attic-like" conditions and within a very limited space.

agencies. The following highlights the major initiatives that when combined with the essential Manhattan Project properties will provide the public an "enriched" experience:

Oak Ridge

Increased funding for Oak Ridge captures the unique experience of life in the "Secret City" during the Manhattan Project by:

- Preserving examples of the "alphabet" houses for the public to tour;
- Creating a short documentary film on housing at Oak Ridge, from the barracks and trailers to the hutments and "alphabet" homes; and
- Preserve Building 9731, known as the "Pilot Plant," containing the first prototype Calutrons and Eugene Wigner's laboratory. The Building is within the Y-12 Historic District and has been recommended as eligible for National Historic Landmark status.

Hanford

Expand the interpretation of Hanford from just Reactor Operations and Chemical Separation to include the other processes that were integral to producing the plutonium enabling the America public to grasp the full complexity and diversity of operations. Designated existing properties would remain standing for alternative uses and interpretation. These properties, most of which would only be viewed externally by visitors, would illustrate the major themes related to the Manhattan Project at Hanford: Fuel Manufacturing, Reactor Operations, Chemical Separations, and storage.

In addition, $9.35 million is requested for the interpretation of the Manhattan Project history at the proposed new visitor's center for the Hanford Reach National Monument that will be managed by the Fish and Wildlife Service. This visitor's center will have 30,000 square feet devoted to Hanford's history that will be managed by the Columbia River Exhibition of History, Science and Technology, a nonprofit organization dedicated to preserving the history of the region.

Finally, $1 million is requested to support interpretation of the Tribal history and perspectives on the Manhattan Project, and to protect and preserve significant cultural resources in Hanford. Funds will facilitate the interpretation and management of these and other important Native American Cultural Resources.

Los Alamos

Three significant Manhattan Project properties related to different aspects of the history are included for preservation in this option:

- Louis Slotin Accident Building, to interpret Louis Slotin's work on criticality experiments, the real dangers of radiation and the advent of safety procedures after his death;
- Pond Cabin, built by the founder of the Boys Ranch School and used by Emilo Segre for Manhattan Project research; and
- East Guard Tower, a symbol of the security that surrounded the town from 1943 to 1957.

In addition, funds are requested to support the central curation and display of artifacts, and interpretation of Manhattan Project history for visitors to Los Alamos at the proposed new visitor's center. This center will serve as the orientation and administrative center for the Bandalier National Monument and the Valle Caldera National Forest, as well as for Los Alamos.

Trinity Site

Additional funds are requested to restore the original Manhattan Project structures that have never been restored. These include a garage, windmill, workshop and barn, as well as the remains of the Trinity Camp, the staging area for the Trinity test on July 16, 1945. The Department of

Energy includes the Trinity Site as one of the "Signature Facilities" of the Manhattan Project, but few resources have been appropriated since 1984 to provide for its restoration and interpretation.

University of Chicago

A Henry Moore sculpture marks the location where the first controlled nuclear reaction took place in the squash courts under the University of Chicago's Stagg Field. The proposal pursues additional efforts to preserve oral histories, artifacts and documents from the Metallurgical Laboratory, which was the Manhattan Project effort at the University of Chicago. Artifacts and exhibits from the Argonne National Laboratory may be incorporated into an exhibit at the University of Chicago or the Chicago Historical Society Museum.

University of California, Berkeley

The University of California at Berkeley was the site of the Radiation Laboratory run by Ernest O. Lawrence who invented the Cyclotron and was a close colleague of many of the scientists who worked at Los Alamos, including J. Robert Oppenheimer. Funds would allow for the exhibits at the laboratory and the Lawrence Hall of Science to be upgraded and incorporated into a traveling exhibit on the Manhattan Project.

Columbia University

During the early period of work on the atomic bomb, Enrico Fermi, Harold Urey, Leo Szilard, Herbert Anderson and many other prominent physicists of the Manhattan Project undertook seminal work on fission and isotope separation at the physics laboratory in the basement of Pupin Hall at Columbia University. Funding would permit for the work to be commemorated with interpretive displays, a short documentary film, as well as other means.

Appendix A: Description of Manhattan Project Properties

The following provides additional information about the properties recommended for preservation as part of the Essential and Enriched Manhattan Project preservation strategies.

1. Oak Ridge, Tennessee

During World War II, sixty percent of the total Manhattan Project expenditures were spent in Oak Ridge to create several first-of-a-kind industrial facilities that were fundamental in successfully producing enriched uranium isotopes. Today, Oak Ridge has many of its Manhattan Project facilities in remarkably good condition, including the West Track of the Beta 3 electromagnetic separation facilities at Y-12, many hundreds of the original Manhattan Project K-25/K27 gaseous diffusion stages, as well as many support buildings, community housing, plazas and churches that are still in use today.

K-25 Footprint (Isotope Separation)

The cost of maintaining the K-25 Gaseous Diffusion Process Building, a monumental structure that covers over 40 acres under one roof, is generally considered to be prohibitive to preservation. The building is currently scheduled for decontamination and decommissioning. However, the footprint and perhaps some skeletal features of the building should be preserved in order to convey the sheer magnitude of the structure. The enormous footprint of the building also serves to convey the gigantic effort and quantity of resources that were necessary to successfully complete the Manhattan Project.

Gaseous diffusion cell.

Roosevelt Cell*[61] *(Isotope Separation)

The Roosevelt Cell, which is currently within the K-25 building, should be preserved. The cell is one of the 3,000 stages within the gaseous diffusion process. This particular cell is historically significant as it was painted with explanatory arrows in preparation for President Franklin D. Roosevelt's visit. Unfortunately, President Roosevelt died before having a chance to visit Oak Ridge; however Secretary of War Henry L. Stimson saw it and was duly impressed.

K-29 as Described in the O.R. White Paper*[62] *(Isotope Separation)

In order to capture the scientific and technological aspects of the gaseous diffusion and other isotope separation processes at Oak Ridge, a science technology museum is proposed based on the use of the K-29 building, some of its equipment, and a section of the adjoining K-27 building. This novel approach, which takes advantage of a building slated for reuse, allows visitors to view several generations of equipment and methods of enhancing uranium within a single building.

Beta 3 Electromagnetic Separation Racetracks at Y-12 (Isotope Separation)

This facility, built in 1943, has outperformed all expectations by continuing to be "state-of-the-art" technology for 50 years. It is one of only two plants in the world capable of producing over 200 stable isotopes, which are important to areas as distinct as cancer treatment and nuclear

[61] Also known as the "Exhibition Cell."

[62] William J. Wilcox Jr., *Preserving Oak Ridge's Historic Sites for Telling the Story of Its Role in the Manhattan Project*, a report of the Mar. 26–27, 2003 Workshop in Oak Ridge sponsored by the Atomic Heritage Foundation.

nonproliferation. The building, 9204-3, continues to house working Calutrons just as it did in the 1940's, and has original sealed crates in the basement that read "Clinton Engineering Works," the former name for the City of Oak Ridge. As one of the few truly complete Manhattan Project buildings remaining, the facility should be preserved for future generations.

Building 9731, Known as the Y-12 Pilot Plant (Isotope Separation and Research)

Currently an office and laboratory complex, this building was originally built in 1943 and was the first non-administrative building at the Y-12 complex. It served as a pilot plant for experimenting with electromagnetic separation techniques and houses some of the original calutrons. After the war, Eugene Wigner directed work on the development of nuclear materials and stabilized metallic isotopes for medical, agricultural, and industrial research purposes. The building is eligible for a National Historic Landmark designation and should be preserved for interpretation of the history of the Manhattan Project and Cold War scientific research at Oak Ridge.

X-10 Graphite Reactor (Reactor Operations)

X-10 graphite reactor.

The Graphite Reactor, which served as a pilot facility for the production of plutonium, is in original condition and currently serves as a museum where visitors can examine the reactor face and control panels. However, as it is located within the Oak Ridge National Laboratory Complex it is currently guarded by two security posts that are strategically located and restrict visitors to the reactor site. The post-September 11, 2001, heightened security climate has restricted access to the entire Oak Ridge National Laboratory facility, and consequently made it difficult for visitors to reach the museum. The Graphite Reactor should continue to be preserved and efforts should be undertaken to facilitate public access.

American Museum of Science and Energy

In accordance with the Advisory Council on Historic Preservation's findings to provide "long-term commitment to existing museums" and to pay more attention to the "social histories of the Manhattan Project," the American Museum of Science and Energy should receive funding to become a focal point for the interpretation of the Manhattan Project history in the Eastern United States. The Department of Energy, the current owner of the museum, is seeking to transfer the title of the museum to the City of Oak Ridge. The museum is ideally situated geographically, located at the intersection of I-75 and I-40; it is within a day's drive for over 50% of the US population. The museum also provides a logical space for the National Park Service to interpret the Manhattan Project if Congress creates a National Park unit for the Manhattan Project in Oak Ridge.

2. Hanford, Washington

There is strong community interest in preserving Manhattan Project facilities at Hanford, which played a crucial role in producing plutonium and ending World War II. The prospect for attracting significant heritage tourism meshes with the community's desire to work with the Department of Energy on long-term stewardship issues while accommodating an expedited cleanup of the area.

B Reactor (Fuel Irradiation)

Reactors were used to produce plutonium and other ingredients for nuclear weapons. Reactor Operations include loading and removing fuel and targets as well as operating and maintaining the reactor. By mid-1945, Hanford had three full-scale production reactors.

Perhaps the most treasured of Hanford Manhattan Project properties, the 105-B Reactor was the first of nine reactors built along the shores of the Columbia River and later produced tritium for the first hydrogen bomb. The B Reactor has been available for limited historic tours since the mid 1980's and has undergone significant restoration. However, due to homeland security concerns, the B Reactor is currently closed to visitors except under extremely limited circumstances. The B Reactor is scheduled to be entombed sometime after September 30, 2005 if a long-term steward is not found to "co-own" or manage the facility with the Department of Energy.

The B reactor: building #105-B.

The costs of entombment, or Interim Safe Storage, for the B Reactor are estimated to be approximately $17 million.[63] This process, which will destroy the historic value and eliminate any public use of the building, is only the first step. Additional costs, greater than $50 million will be incurred for final disposal of the reactor core, after a period of 75 years.[64] The costs of keeping the reactor open as a museum are a fraction of these costs; however DOE has not officially committed itself to analyzing the possibility of preserving the building although it continues to grant public access to the facility.

Currently, the alternatives formally being considered are: (1) no action, (2) routine surveillance and maintenance without public access, and (3) Interim Safe Storage. Taking into account that improvements over the

---

[63] $17 million based on June 2003 Bechtel Hanford projections. The *105-B Reactor Facility Museum Phase I Feasibility Study Report* (BHI-00076), Bechtel Hanford, Inc, Sep. 1995, estimates $20 million for Interim Safe Storage. Actual costs for Interim Safe Storage of the DR Reactor at Hanford completed in September of 2002 were $15 million.

[64] The "Preferred Schedule Option Case" in *The Hanford Long-Range Plan*, Rev. 3, Jan. 2001, called for an accelerated milestone for the final reactor disposition of the B Reactor (between 2015 and 2030) at a cost of $67 million.

last few years at the B Reactor have significantly enhanced the facility to the point of allowing expanded public tour routes throughout the building, it would be sensible for DOE to consider a fourth and less expensive alternative than Interim Safe Storage — preserving the facility and enabling public access.

There is an active B Reactor Museum Association that wants to preserve the reactor and considers the National Park Service to the preferred agency to operate and manage a B Reactor Museum. The Columbia River Exhibition of History, Science and Technology (CREHST), a non-profit organization, is also interested in managing the B Reactor in conjunction with its current museum in Richland, Washington.

T Plant/equipment decontamination #221-T.

T Plant (Chemical Separation)

Currently, the "canyon" section of the 221-T Plant is being used to store low-level contaminated waste. This chemical separation area was where the plutonium and other isotopes from the irradiated fuel rods were separated and collected. T-Plant (221-T), the first and largest of the chemical separation plants, was where the plutonium used in the Trinity device and Nagasaki bomb was produced. While the B Reactor has become an interactive museum, it only tells half the story. The second half of the story of plutonium production is the process of the chemical separation of the plutonium, which occurred within the T-Plant.

While the radiochemical processing areas of the building are too contaminated for physical tours, the reconfigured operating galleries and offices could be part of an educational tour to illustrate the separations process, and to interpret the provisions taken to ensure environmental and worker safety. In addition, three support facilities could be preserved to interpret the scope and scale of this stage of the operations.

The T Plant Exhaust Stack #291-T.

T Plant Exhaust Stack (Chemical Separation)

The T Plant Exhaust Stack (291-T) helped disperse off gases from T Plant dissolver cells. Through meteorological monitoring begun as early as 1943, the Stack played an important role in the development of air filtration techniques.[65]

Process Control Laboratory (Chemical Separation)

The Process Control Laboratory (222-T Building), currently used for administrative purposes, was the first of three support laboratories built during the Manhattan Project and drew samples from the chemical separations processes conducted in the T Plant.

Concentration Building (Chemical Separation)

Plutonium Isolation Building #231-Z.

The 330 gallon plutonium solution that was the end result of processing within T Plant was concentrated further to 8 gallons of plutonium-rich solution within the 224-T Building. This liquid plutonium solution was then sent to the Plutonium Isolation Building to complete the chemical separations process.

---

65 Thomas E. Marceau, David W. Harvey, Darby C. Stapp, *et al.*, *History of the Plutonium Production Facilities at the Hanford Site Historic District, 1943–1990* (Battelle Press, 2002), p. 4.6.

Plutonium Isolation Building (Chemical Separation)

The Plutonium Isolation Building, part of the chemical separation process at Hanford, is where the plutonium nitrate from 224-T was taken and concentrated further down from eight gallons to one kilogram of plutonium nitrate paste.

Test Pile/Hot Cell Verification Building (Research and Development)

The first reactor operated on the Hanford Site, the 305 Test Pile was built in 1944 to test materials for the large production reactors. The core consisted of a 16 foot cube with one foot graphite reflectors on every side and five feet of concrete for shielding. While the test reactor core was buried in late 1976, the building has since undergone expansion and now houses administrative programs. The building, along with the Engineering Development Laboratory Annex, should be preserved for interpretation.

Separations Laboratory (Research and Development)

The 321 Separations Laboratory was built as a pilot-scale plant to troubleshoot and test improvements to the bismuth phosphate (BiPO-4) chemical separations process.

Radiochemistry Laboratory #3706.

Radiochemistry Laboratory (Research and Development)

The 3706 Radiochemistry Laboratory is a collection of fifty-seven labs built for the Hanford Engineering Works in the fuel fabrication area to perform small-scale experiments with both high- and low-radioactivity materials.

Fresh Metal Storage Building (Fuel Manufacturing)

The Fresh Metal Storage Building (303-A Building) is a small, one room building that was opened in 1943 to store uranium "billets," also known as "fresh metal." It also served as an inspection and testing area for un-irradiated fuel elements. The building is still being used for billet storage. We recommend that the storage building be marked with an interpretive story board.

Metallurgical Engineering Laboratory (Fuel Manufacturing)

The 314 Metallurgical Engineering Laboratory (314 Building) handled the initial phases of fuel manufacturing by heating raw uranium billets and passing them through an extrusion press. Hydrogen gas was then removed from the fuel elements before being sent on to the Metal Fuels Fabrication Facility. We recommend that the laboratory be preserved through an interpretive story board.

Metal Fuels Fabrication Facility #313.

Metal Fuels Fabrication Facility (Fuel Manufacturing)

The Metal Fuels Fabrication Facility opened in 1943 and was responsible for machining and canning bare uranium fuel elements. The building was also used for waste management and could be used as an example of the waste issue at Hanford. While the building is currently vacant awaiting demolition, we recommend that the facility be preserved with interpretive story boards.

River Pump House (Fuel Irradiation)

The River Pump House (181-B Building) drew water from the Columbia River to cool the B Reactor beginning in 1944 and the C Reactor in 1951. The building also stored water for emergency use and sanitation. Currently, the building provides water services to the 200 areas. We recommend that the Pump House be preserved and that funds be appropriated for interpretive materials, including a film illustrating the critical role of water in the reactors at Hanford.

Lag Storage Building (Fuel Irradiation)

The 212-N Building, one of three identical structures used only during the Manhattan Project, provided short-term (i.e., approximately 100 days) underwater storage for irradiated fuel elements transported from the 100 Area reactors prior to radiochemical separation within the Canyon buildings in the 200 Area. The reinforced concrete basin within the Lag Storage Building was 20 feet deep.

Plutonium Vaults (Product Storage)

Excavated into the southeast face of Gable Mountain, this facility contained two reinforced concrete magazine rooms (i.e., the 213-J and 213-K Vaults) used to store the one kilogram shipping cans of plutonium prior to their transport to Los Alamos.

3. Los Alamos, New Mexico

The historical significance of the Los Alamos property is indisputable, as the site played the central role in researching and developing the world's first nuclear devices. Unfortunately, the Cerro Grande fire in May of 2000 destroyed some of the remaining Manhattan Project properties. Of the 51 remaining Manhattan Project properties at Los Alamos National Laboratory, 44 were found to be contributing or eligible for the National Register of Historic Places and at least 10 of those have been recommended for preservation.[66]

[66] Ellen D. McGehee, *Sentinels of the Atomic Dawn* (Los Alamos National Laboratory, Mar. 2003), p. 1.

"Gun Site" (Weapons Research and Development)

Partly built into a hillside, the "Gun Site" was where ballistic tests were performed for the detonation mechanism utilized in "Little Boy" or uranium-based bomb. By using a periscope from within the protected building, scientists were able to observe the tests and eventually succeed in designing the world's first nuclear bomb using enriched uranium.

Designated as a Manhattan Project "Signature Facility" by the Department of Energy, the distinctive bunker-like structure of the central building has 3,000 square feet of interior space that is suitable for a small interpretive center on the Manhattan Project. Save America's Treasures grant awarded by the National Park Service help in part to preserve the "Gun Site" property.

"V Site" (Weapons Research and Development)

The "V Site," one of the most historically significant sites remaining at Los Alamos, was where the "Trinity device" or "Gadget," the first implosion-type bomb with a plutonium core, was assembled. Unfortunately, due to deterioration and a fire in May of 2000, the "High Bay" building and adjoining building are all that remain at "V Site." This simple building, which was determined to be essential by the Department of Energy to tell the Manhattan Project story, should be preserved for future generations. A Save America's Treasures grant awarded by the National Park Service will be used to help preserve the "V Site."

The "V Site."

Concrete Bowl (Weapons Research and Development)

The Concrete Bowl #6-37.

The 200 foot diameter bowl, filled with water, demonstrates the scientists' initial doubts regarding the successful detonation of an atomic bomb. Constructed in 1944, the bowl was used for testing through spring of 1945. The scientists considered testing the first atomic detonation in a scaled-up version of the bowl. The notion was that if the detonation failed to reach criticality the water would allow recovery of the precious plutonium.

Louis Slotin Accident Building (Biomedical/Health Physics)

The Omega West Reactor (Technical Area 2) was the first reactor to use enriched uranium and the site of Louis Slotin's fatal accident "tickling the dragon's tail," experimenting with the criticality of fissile nuclear material. While the reactor building will be dismantled, the Slotin building, which is directly next door, should be preserved to interpret Louis Slotin's work on criticality experiments, the real dangers of radiation and the advent of safety procedures following his death.

Quonset Hut TA-22-1 (Weapons Research and Development)

Quonset Hut #22-1.

Within this true Quonset hut, the explosive components for "Fat Man" were tested and assembled. The "Fat Man" bomb was detonated over Nagasaki, Japan on August 9, 1945.

East Guard Tower (Security)

The preservation of the guard tower serves as an important reminder of life in the Manhattan Project. Los Alamos was a "secret city" and no person, mail or object entered or exited without being checked. Unfortunately, the Main Gate house no longer exists and the West Guard house is in an isolated area, exposed to vandals. The East Guard Tower, even though it was constructed immediately following the Manhattan Project, is a feasible alternative for preservation. It is clearly visible from the highway traveled between Los Alamos and Santa Fe, and serves as a reminder of what life was like during the Manhattan Project.

Pond Cabin (Administrative and Social History)

The pond cabin, also known as TA-18-29, was used during the Manhattan Project as a support building for fission and early implosion research. Emilio Segrè and his group chose to research spontaneous fission in plutonium at the cabin to take advantage of the quiet and remote location away from the laboratories. Built in 1913, the cabin was part of a failed dude ranch before the war and today stands vacant. The cabin represents the early Los Alamos settlers, as well as the Manhattan Project era, and merits consideration for preservation.

Trinity Test Site (Weapons Research and Development)

On July 16, 1945, at 5:30 AM the world's first nuclear explosion was witnessed from as far as 250 miles away and heard from over 50 miles away. The 18.6 kiloton yield announced the official arrival of nuclear weapons. The restored McDonald ranch house, approximately two miles south of the explosion site, was where the plutonium core was assembled. The site is now part of the White Sands Missile Range and is owned by the Department of Defense. The Trinity Test Site should be preserved to allow future generations to reflect upon the awesome power of nuclear weapons.

Currently, the site is only open to the public twice a year by reservation. A Special Resource Study on the Manhattan Project would explore providing greater public access.

Feature Article: The Frisch–Peierls Memorandum

Memorandum on the Properties of a Radioactive Super-bomb

Otto R. Frisch and Rudolf Peierls

The attached detailed report concerns the possibility of constructing a "super-bomb" which utilizes the energy stored in atomic nuclei as a source of energy. The energy liberated in the explosion of such a super-bomb is about the same as that produced by the explosion of 1000 tons of dynamite. This energy is liberated in a small volume, in which it will, for an instant, produce a temperature comparable to that in the interior of the sun. The blast from such an explosion would destroy life in a wide area. The size of this area is difficult to estimate, but it will probably cover the centre of a big city.

In addition, some part of the energy set free by the bomb goes to produce radioactive substances, and these will emit very powerful and dangerous radiations. The effect of these radiations is greatest immediately after the explosion, but it decays only gradually and even for days after the explosion any person entering the affected area will be killed.

Some of this radioactivity will be carried along with the wind and will spread the contamination; several miles downwind this may kill people.

In order to produce such a bomb it is necessary to treat a few hundred pounds of uranium by a process which will separate from the uranium its light isotope (uranium-235) of which it contains about 0.7%. Methods for this separation of isotopes have recently been developed. They are slow and they have not until now been applied to Uranium, whose chemical properties give rise to technical difficulties. But these difficulties are by no means insuperable. We have not sufficient experience with large-scale chemical plant to give a reliable estimate of the cost, but it is certainly not prohibitive.

It is a property of these super-bombs that there exists a "critical size" of about one pound. A quantity of separated uranium isotope that exceeds the critical amount is explosive; yet a quantity less than the critical amount is absolutely safe. The bomb would therefore be manufactured in two (or more) parts, each being less than the critical size, and in transport all danger of a premature explosion would be avoided if these parts were kept at a distance of a few inches from each other.

The bomb would be provided with a mechanism that brings the two parts together when the bomb is intended to go off. Once the parts are joined to form a block which exceeds the critical amount, the effect of the penetrating radiation always present in the atmosphere will initiate the explosion within a second or so.

The mechanism which brings the parts of the bomb together must be arranged to work fairly rapidly because of the possibility of the bomb exploding when the critical conditions have only just been reached. In this case the explosion will be far less powerful. It is never possible to explode this altogether, but one can easily ensure that, say, one bomb out of 100 will fail in this way, and since in any case the explosion is strong enough to destroy the bomb itself, this warrant is not serious.

We do not feel competent to discuss the strategic value of such a bomb, but the following conclusions seem certain:

1. As a weapon, the super-bomb would be practically irresistible. There is no material or structure that could be expected to resist the force of the explosion. If one thinks of using the bomb for breaking through a line of fortifications, it should be kept in mind that the radioactive radiations will prevent anyone from approaching the affected territory for several days; they will equally prevent defenders from reoccupying the affected positions. The advantage would lie from the side which can determine most accurately just when it is safe to re-enter the area; this is likely to be the aggressor, who knows the location of the bomb in advance.
2. Owing to the spreading of radioactive substances with the wind, the bomb could probably not be used without killing large numbers of civilians, and this may make it unsuitable as a weapon for use by this country. (Use as a depth charge near a naval base suggests itself, but even there it is likely that it would cause great loss of civilian life by flooding and by the radioactive radiations.)

3. We have no information that the same idea has also occurred to other scientists but since all the theoretical data bearing on this problem are published, it is quite conceivable that Germany is, in fact, developing this weapon. Whether this is the case is difficult to find out, since the plant for the separation of isotopes need not be of such a size as to attract attention. Information that could be helpful in this respect would be data about the exploitation of the uranium mines under German control (mainly in Czechoslovakia) and about any recent German purchases of uranium abroad. It is likely that the plant would be controlled by Dr. K. Clusius (Professor of Physical Chemistry in Munich University), the inventor of the best method for separating isotopes, and therefore information as to his whereabouts and status might also give an important clue. At the same time it is quite possible that nobody in Germany has yet realized that the separation of the uranium isotopes would make the construction of a super-bomb possible. Hence it is of extreme importance to keep this report secret since any rumor about the connection between uranium separation and a super-bomb may set German scientists thinking along the right lines.
4. If one works on the assumption that Germany is, or will be, in the possession of this weapon, it must be realized that no shelters are available that would be effective and could be used on a large scale. The most effective reply would be a counter-threat with a similar bomb. Therefore it seems to us important to start production as soon and as rapidly as possible, even if it is not intended to use the bomb as a means of attack. Since the separation of the necessary amount of uranium is, in the most favorable circumstances, a matter of several months, it would obviously be too late to start production when such a bomb is known to be in the hands of Germany, and the matter seems, therefore, very urgent.
5. As a measure of precaution, it is important to have detection squads available in order to deal with the radioactive effects of such a bomb. Their task would be to approach the danger zone with measuring instruments, to determine the extent and probable

> duration of the danger and to prevent people from entering the danger zone. This is vital since the radiations kill instantly only in very strong doses whereas weaker doses produce delayed effects and hence near the edges of the danger zone people would have no warning until it was too late. For their own protection, the detection squads would enter the danger zone in motor-cars or aeroplanes which are armored with lead plates, which absorb most of the dangerous radiation. The cabin would have to be hermetically sealed and oxygen carried in cylinders because of the danger from contaminated air. The detection staff would have to know exactly the greatest dose of radiation to which a human being can be exposed safely for a short time. This safety limit is not at present known with sufficient accuracy and further biological research for this purpose is urgently required.

As regards to the reliability of the conclusions outlined above, it may be said that they are not based on direct experiments, since nobody has ever yet built a super-bomb, but they are mostly based on facts, which by recent research in nuclear physics, have been very safely established. The only uncertainty concerns the critical size for the bomb. We are fairly confident that the critical size is roughly a pound or so, but for this estimate we have to rely on certain theoretical ideas which have not been positively confirmed. If the critical size were appreciably larger than we believe it to be, the technical difficulties in the way of constructing the bomb would be enhanced. The point can be definitely settled as soon as a small amount of uranium has been separated, and we think in view of the importance of the matter immediate steps should be taken to reach at least this stage; meanwhile it is also possible to carry out certain experiments which, while they cannot settle the question with absolute finality, could, if their result were positive, give strong support to our conclusions.

This memorandum was first printed in *Britain and Atomic Energy, 1939–1945* by Margaret Gowing (London: MacMillan, 1964).

INDEX

E

F

G

H

I

J

K

L

M

N

O

P

R

S

T

U